THE
HoW
?
BOOK SERIES

Everything in this book is factual and scientific. The most outlandish claims and stories are simply the most true.

Based on the design and template of Mark Wasserman and Irene Ng of Plinko, this volume was designed by Eliana Stein.

Manufactured in Singapore.

10 9 8 7 6 5 4 3 2 1

Library of Congress Cataloging-in-Publication Data is available.

Published by the HOW Series, Dedicated to the Exploration and Dissemination of Unbelievable Brilliance.

Photographs of Dr. and Mr. Haggis-On-Whey by Meiko Takechi Arquillos.
Cover illustration by Michael Kupperman.
Photograph of possum family by Susan Remele of Dallas, Texas.
Photographs of popcorn by John Linwood.

This book is dedicated to Adams, Russ; Adee, George*; Alexander, Frederik*; Allison, Willmer*; Alonso, Manuel*; Anderson, Malcolm; Ashe, Arthur*; Atkinson, Juliette*; Austin, Henry*; Austin, Tracy; B; Baker, Lawrence*; Barger-Wallach, Maud*; Becker, Boris; Behr, Karl*; Betz (Addie), Pauline; Bolton, Nancy*; Borg, Bjorn; Borotra, Jean*; Bromwich, John*; Brookes, Sir Norman*; Brough (Clapp), Louise; Browne, Mary*; Brugnon, Jacques*; Buchholz, Earl; Budge, John*; Bueno, Maria; C; Cahill, Mabel*; Campbell, Oliver*; Casals, Rosie (Rosemary); Chace, Malcolm*; Chang, Michael; Chatrier, Philippe*; Cheney, Dorothy; Clark, Clarence*; Clark, Joseph*; Clerici, Gianni; Clothier, William*; Cochet, Henri*; Collins, Arthur; Connolly Brinker, Maureen*; Connors, James; Cooper, Ashley; Courier, Jim; Court Smith, Margaret; Crawford, Jack*; Cullman, Joseph*; D; Danzig, Allison*; David, Herman*; Davidson, Sven*; Davis, Dwight*; Dod, Charlotte*; Doeg, John*; Doherty, Laurence*; Doherty, Reginald*; Douglass Lambert Chambers, Dorothea*; Drobny, Jaroslav*; duPont, Margaret; Durr, Francoise; Dwight, James*; E; Edberg, Stefan; Emerson, Roy; Etchebaster, Pierre*; Evert, Christine; F; Falkenburg, Robert; Farquhar, Marion*; Fraser, Neale; Fry-Irvin, Shirley; G; Garland, Charles*; Gibson, Althea*; Gonzalez, Richard*; Goolagong Cawley, Evonne; Gore, Arthur*; Graf, Steffi; Grant, Bryan*; Gray, David*; Griffin, Clarence*; Gustav V, King of Sweden*; H; Hackett, Harold*; Hansell Allderdice, Ellen*; Hard, Darlene; Hart, Doris; Haydon Jones, Adrianne; Heldman, Gladys*; Hester, William*; Hewitt, Robert; Hoad, Lewis*; Hopman, Harry*; Hotchkiss Wightman, Hazel*; Hovey, Frederick*; Hunt, Joseph*; Hunt, Lamar*; Hunter, Francis*; J; Jacobs, Helen*; Johnston, William*; Jones, Perry*; K; Kelleher, Robert; King, Billie Jean; Kodes, Jan; Kozeluh, Karel*; Kramer, Jack; L; Lacoste, Jean*; Laney, Albert*; Larned, William*; Larsen, Arthur; Laver, Rod; Lawford, Herbert*; Lendl, Ivan; Lenglen, Suzanne*; Lott, George*; M; Mako, Constantine; Mallory, Anna*; Mandlikova, Hana; Marble, Alice*; Martin, Alastair; Martin, William*; Maskell, Dan*; Mathieu, Simonne*; McCormack, Mark*; McEnroe, John; McGregor, Ken*; McKane Godfree, Kathleen*; McKinley, Charles*; McLoughlin, Maurice*; McMillan, Frew; McNeill, William; Moore, Elisabeth*; Mortimer Barett, Florence; Mulloy, Gardnar; Murray, Robert*; Myrick, Julian*; N; Nastase, Ilie; Navratilova, Martina; Newcombe, John; Nielsen, Arthur*; Noah, Yannick; Novotna, Jana; Nusslein, Hans*; Nuthall Shoemaker, Betty*; O; Olmedo, Alejandro; Osuna, Rafael*; Outerbridge, Mary*; P; Palfrey Fayban Cooke Danzig, Sarah*; Parker, Frank*; Patterson, Gerald*; Patty, John; Pell, Theodore*; Perry, Frederick*; Pettitt, Thomas*; Pietrangeli, Nicola; Q; Quist, Adrian*; R; Rafter, Patrick; Ralston, Richard; Renshaw, James*; Renshaw, William*; Richards, Vincent; Richey, Nancy; Riggs, Robert; Roche, Anthony; Roosevelt, Ellen*; Rose, Mervyn; Rosewall, Kenneth; Round Little, Dorothy*; Ryan, Elizabeth*; S; Sabatini, Gabriela; Sampras, Pete; Sanchez-Vicario, Arantxa; Santana, Manuel; Savitt, Richard; Schroeder, Fredrick*; Scott, Eugene*; Sears, Eleonora*; Sears, Richard*; Sedgman, Frank; Segura Caano, Francisco; Seixas, Elias; Shields, Francis*; Shoemaker, Betty May*; Shriver, Pam; Slocum, Henry*; Smith, Stanley; Stolle, Fredrick; Sutton Bundy, May*; T; Talbert, William*; Tilden, William*; Tingay, Lance; Tinling, Cuthbert; Tobin, Brian; Townsend Toulmin, Bertha*; Trabert, Tony; Turner Bowrey, Lesley; V; Van Alen, James*; Van Ryn, John*; Vilas, Guillermo; Vines, Henry*; von Cramm, Gottfried*; W; Wade, Sarah; Wagner, Marie*; Ward, Holcombe*; Washburn, Watson*; Whitman, Malcolm*; Wilander, Mats; Wilding, Anthony*; Williams, Richard*; Wills Moody Roark, Helen*; Wingfield, Walter*; Wood, Sidney; Wrenn, Robert*; Wright, Beals*.

**Deceased*

www.haggis-on-whey.com
www.mcsweeneys.net

ISBN 978-1-934781-21-0

WELCOME! TO

DR. AND MR. DORIS HAGGIS-ON-WHEY'S WORLD OF UNBELIEVABLE BRILLIANCE

This is the fourth in what will soon enough be a series of 176 books about everything worth knowing on this planet or any other. These books, all classics of their kind, have been and will continue to be written by me, Dr. Doris Haggis-On-Whey, the foremost person of knowledge among all people. The person pictured with me is Benny, my husband and formerly ward of the village of Crumpets-Under-Kilt, where we now reside. Benny now lives in the shed so as not to interfere with my work or meals.

On the next few pages, find highlights from the previous three H-O-W books.

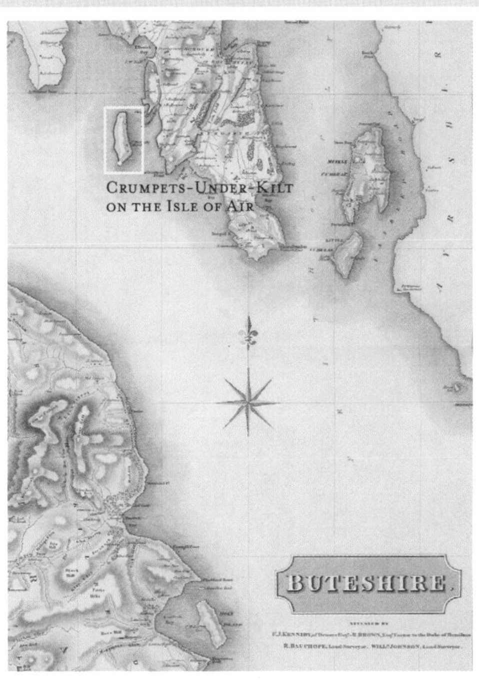

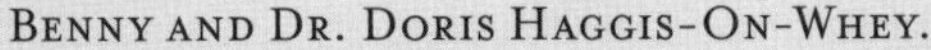

BENNY AND DR. DORIS HAGGIS-ON-WHEY.

THE SHED.

VOLUME 1

Giraffes? Giraffes!

This book, a stunning retort to all Darwin posited, proves that this improbable animal came not from any horse-like beast but from space, via conveyor. I also clarify the role of giraffes in what we see in mirrors, and I go into helpful detail about the relationship between giraffes and the better cities of Indiana.

THE ORIGIN OF GIRAFFES

Giraffes first came to this planet nearly five-hundred thousand years ago, on a conveyor belt. No one is sure where the conveyor belt came from, because the pieces of the conveyor belt recovered for scientific study — in 1973, in Middleton, New Jersey, by Arni Arnarsson, originally from Iceland — are being hidden from the authors, Dr. and Mr. Haggis-On-Whey, by governmental stooges. The authors, however, know that this conveyor existed, because they have a very good hunch about it, and because Arni Arnarsson was pretty sure, too. The conveyor is believed by the authors, one of whom is a trained scientist, to have originated on Neptune. This is the opinion of the authors because most scientists believe Neptune, because of its unique gaseous makeup and its green color, is the most likely planet to be inhabited by giraffes who could build conveyors.

ARNARSSON

EARTH
SATURN
VENUS
MERCURY
URANUS
PLUTO
MARS
NEPTUNE
JUPITER

SIZE OF PLANETS, RELATIVE TO EACH OTHER

EARTH
MARS
SATURN
VENUS
MERCURY
NEPTUNE (REPRISE)
PLUTO
URANUS
NEPTUNE
MARS

PLANETS SHOWN IN PROPORTION TO THE ODDS THAT IT TRULY IS THE ORIGIN OF ALL GIRAFFES

* A CERTAIN SPECIES OF MALE GIRAFFES MAY BE FROM MARS

HOW MANY GIRAFFES CAME HERE?

7 million	
8 million	¡8.5 MILLION!

At first, there were only 7 million giraffes on Earth, but they quickly multiplied until there were 8.5 million giraffes.

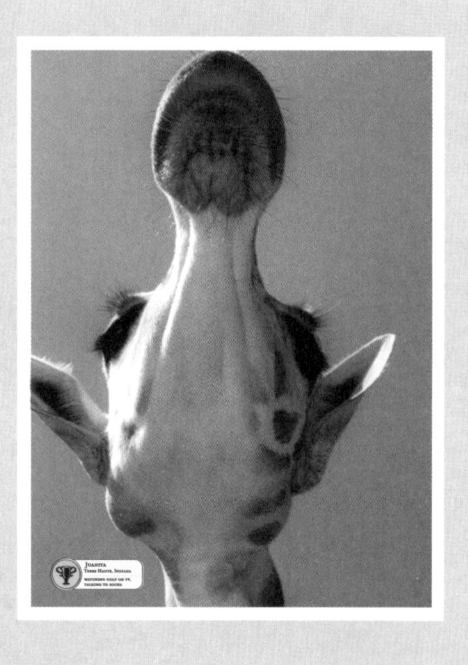

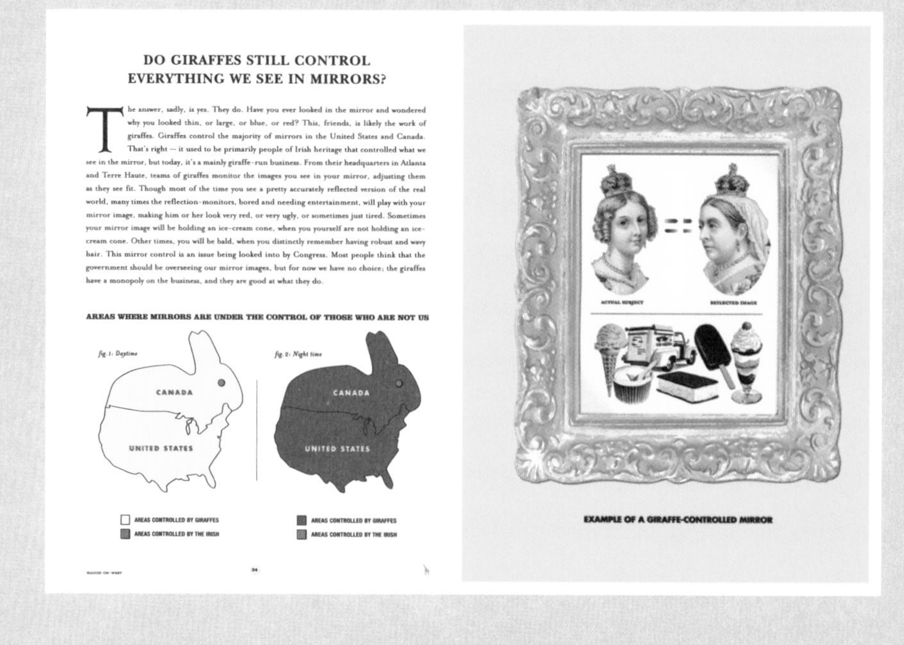

DO GIRAFFES STILL CONTROL EVERYTHING WE SEE IN MIRRORS?

The answer, sadly, is yes. They do. Have you ever looked in the mirror and wondered why you looked thin, or large, or blue, or red? This, friends, is likely the work of giraffes. Giraffes control the majority of mirrors in the United States and Canada. That's right — it used to be primarily people of Irish heritage that controlled what we see in the mirror, but today, it's a mainly giraffe-run business. From their headquarters in Atlanta and Terre Haute, teams of giraffes monitor the images you see in your mirror, adjusting them as they see fit. Though most of the time you see a pretty accurately reflected version of the real world, many times the reflection-monitors, bored and needing entertainment, will play with your mirror image, making him or her look very red, or very ugly, or sometimes just tired. Sometimes your mirror image will be holding an ice-cream cone, when you yourself are not holding an ice-cream cone. Other times, you will be bald, when you distinctly remember having robust and wavy hair. This mirror control is an issue being looked into by Congress. Most people think that the government should be overseeing our mirror images, but for now we have no choice; the giraffes have a monopoly on the business, and they are good at what they do.

AREAS WHERE MIRRORS ARE UNDER THE CONTROL OF THOSE WHO ARE NOT US

fig. 1: Daytime

fig. 2: Night time

CANADA
UNITED STATES

AREAS CONTROLLED BY GIRAFFES
AREAS CONTROLLED BY THE IRISH

EXAMPLE OF A GIRAFFE-CONTROLLED MIRROR

VOLUME 2

Your Disgusting Head

This volume, sometimes considered lesser than my other books, is lesser than my other books. It is about the substances that fill your face and skull, and has been misunderstood and attacked by my "colleagues" at the *JAMA* and *Lancet*. They will get their comeuppance when I replace my pellet rifle.

THE JUICES, OILS, GOOS, GUNKS, OOZES, AND SAPS THAT MAKE UP YOUR HEAD ARE OF MANY TYPES, AND OCCUPY MANY DIFFERENT PLACES IN YOUR HEAD. HERE IS A GUIDE, NOW AND FOR ALWAYS, TO:

THE SICKENING FLUIDS THAT FILL YOUR SKULL

There are over 1,100 known fluids in your skull and face, and they are all uniformly repulsive and most of them smell like soap when it has been dropped in tar. For centuries, scientists have attempted to categorize these fluids, to no avail. Most recently, a team from Holland tried, but these Hollandnese scientists were too tall to fit into their labs, and gave up. What makes it so hard to catalog all of these liquids and jellies? First of all, they are very slippery. Secondly, some of them are invisible to the naked eye. Thirdly, have you ever had a headache caused by a husband who is trying to learn to play the cello but thinks he can do so without a bow? Have you had a husband, named Benny, who just sort of bangs on the cello like it's a drum or kitten? It can be very distracting when you're trying to make sense of the putrid crud that fills our heads.

FLUID	NICKNAME	LOCATION	FUNCTION
Snot	"Snotty"	Nose, back of throat	None
Saliva	"Spitty"	Mouth	None
Drool	"Drooly"	Mouth, couch, bed	Public humiliation
Blood	"Red"	Brain, nose, ears	None
Boogers	"Boogeyman"	Nose, fingers	For hobbyists
Wax	"Lefty"	Ears	Smells interesting
Pus	"Pus-Man"	Nose, ears, brain	None
Goop	"MC G-Star"	Back of throat	Makes voice sound froggy
Spittle	"Spittle-izzle"	Corner of mouth	Impresses the ladies
Ooze	"Ms. Oozey"	Brain	None known
Gunk	"Dr. Gunkmeister"	Brain	Helps with memories of accidents with food
Plasma	"J. Plasma Funkmasta"	Brain, neck, face (under)	Tells jokes, some are funny
Sap	"Johnny Sap"	Upper nose, lower ear	None

COMPOSITE PORTRAIT OF ALL U.S. PRESIDENTS

WHOSE MOUTH DO YOU HAVE?

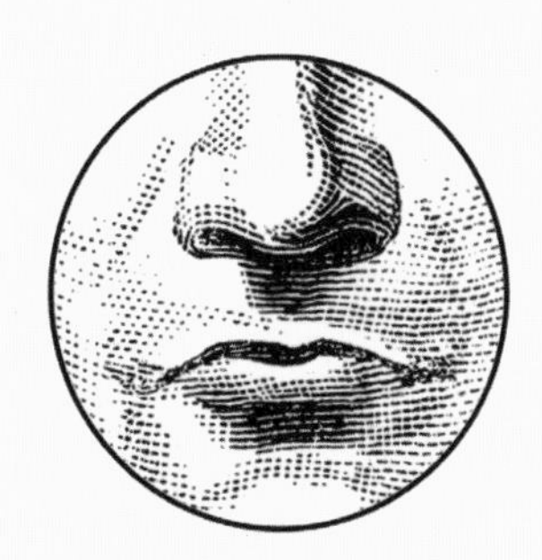

Your father's

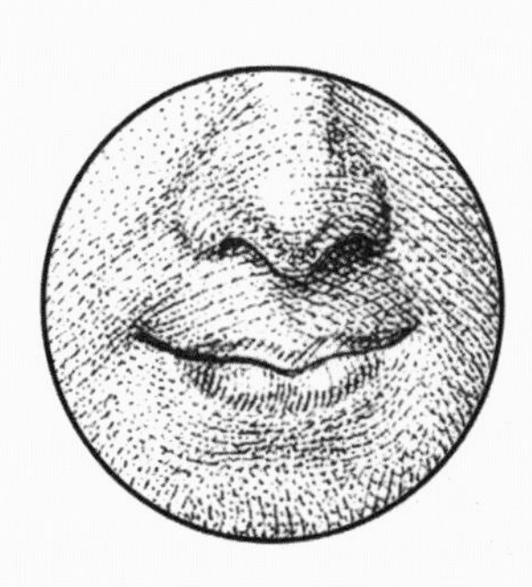

Your mother's

Chester A. Arthur's

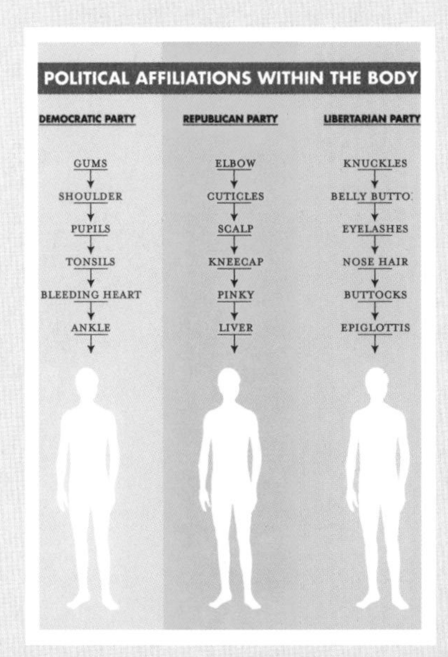

VOLUME 3

Animals of the Ocean, in Particular the Giant Squid

This is considered by most the definitive book on the sea, on the animals in the sea, and on animals in general and biology *in toto*. It covers the most popular baby names for all manner of marine life, it blows the lid off shark religion, and explains why rank-and-file giant squid object to the work of E.M. Forster.

MOST POPULAR BABY NAMES FOR FIFTY SELDOM SEEN SEA ANIMALS

ANIMAL	PICTURE	BOYS	GIRLS	ANIMAL	PICTURE	BOYS	GIRLS
Sea Horse		Michael	Alexis	Halibut		Brady	India
Stingray		Michael, Keith	Alexis	Hatchetfish		Preston	Raven
Ribbonfish		Tyler	Olivia	Catfish		Dustin	Ruby
Swordfish		Dylan	Mackensie/Makenzie	Rabbitfish		Chandler	Dakota
John Dory		Brandon	Jordan	Mussel		Taylor	Cheyenne
Mackerel		Brandon	Trinity	Pufferfish		Cody	Genesis
Silverside		Austin	Sierra	Little Post-Horn Squid		Cody	Brooklyn
Thresher Shark		Caleb	Dakota	Octopus		Damian	Judaica
Moonfish		Cody	Bailey	Fu Manchu		Dawson	Asia
Spiny-headed Worm		Noah, Cody	Miranda	Giant Squid		Andre	Guantanamo/Guantanama
Sperm Whale		Mason	Cheyenne	Eel		Shane	Mercedes
Flatworm		Elijah	Sabrina	Sea Urchin		Derek	Dakota
Killer Whale		Kyle	Angel	Squid		Travis	Hope
Tongue Worm		Elijah	Kiara	Snipe Eel		Noah	Condoleeza
Snakehead		Cameron	Kennedy	Dirty Phish Fan		Cody	Dakota
Gastrotrich		Trevor	Chloe	Lobster		Cody	Cheyenne
Carp		Cody	Autumn	Big-Headed Rattail		Chestnut	Richard
Jellyfish		Aidan	Skylar/Skyler	Wonder-lamp Squid		Brady	Apple
Crab		Dakota	Destiny	Oarfish		Jordache	Tangelo
Dolphin		Bryce	Aaliyah/Aliyah	Batfish		Cody	Dakota
Salmon		Colby	Tatiana	Lazy Squid		Parker	South Dakota
Basking Shark		Alec	Ariel	Pelican Eel		Henry/Henri	Addison
Ray's Bream		Colton	Delaney	Devilfish		Michael	State
Prawn		Omar	Dana	Starfish		Elijah	Wacker
Flying Fish		Jose	Reagan	Lanternfish		Frodo	Michigan

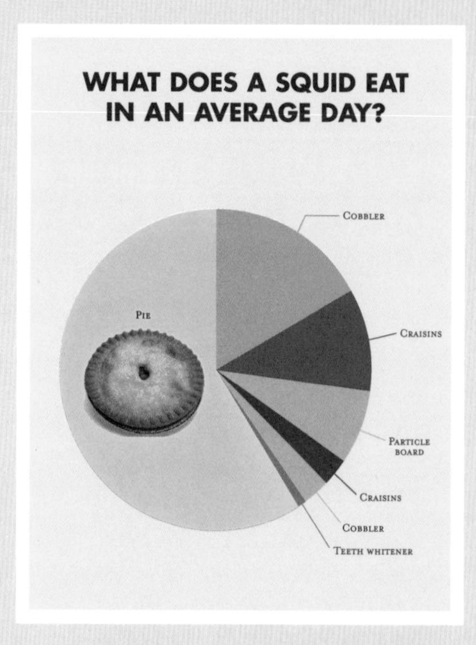

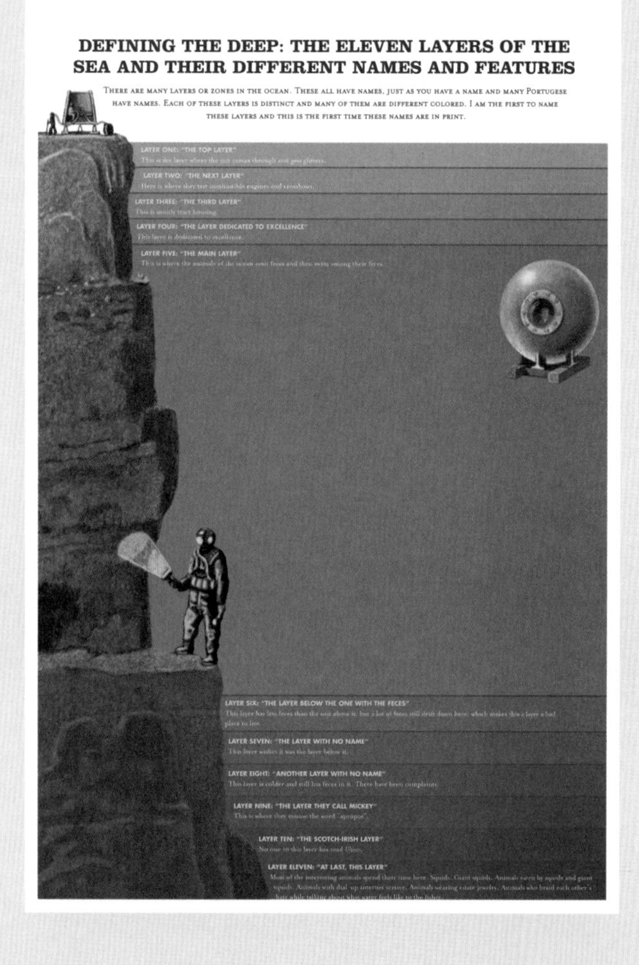

FROM THE AUTHOR

A Typical Doris Day

Many of you readers have figured out how to use basic tools and even to write letters. In these letters many of you have asked for an explanation of how I spend my days. I laughed for a long time at your impertinence. And then I had a bad idea, which was to answer your question. I hope this is a first step in severing our relationship forever.

In 24 Hours

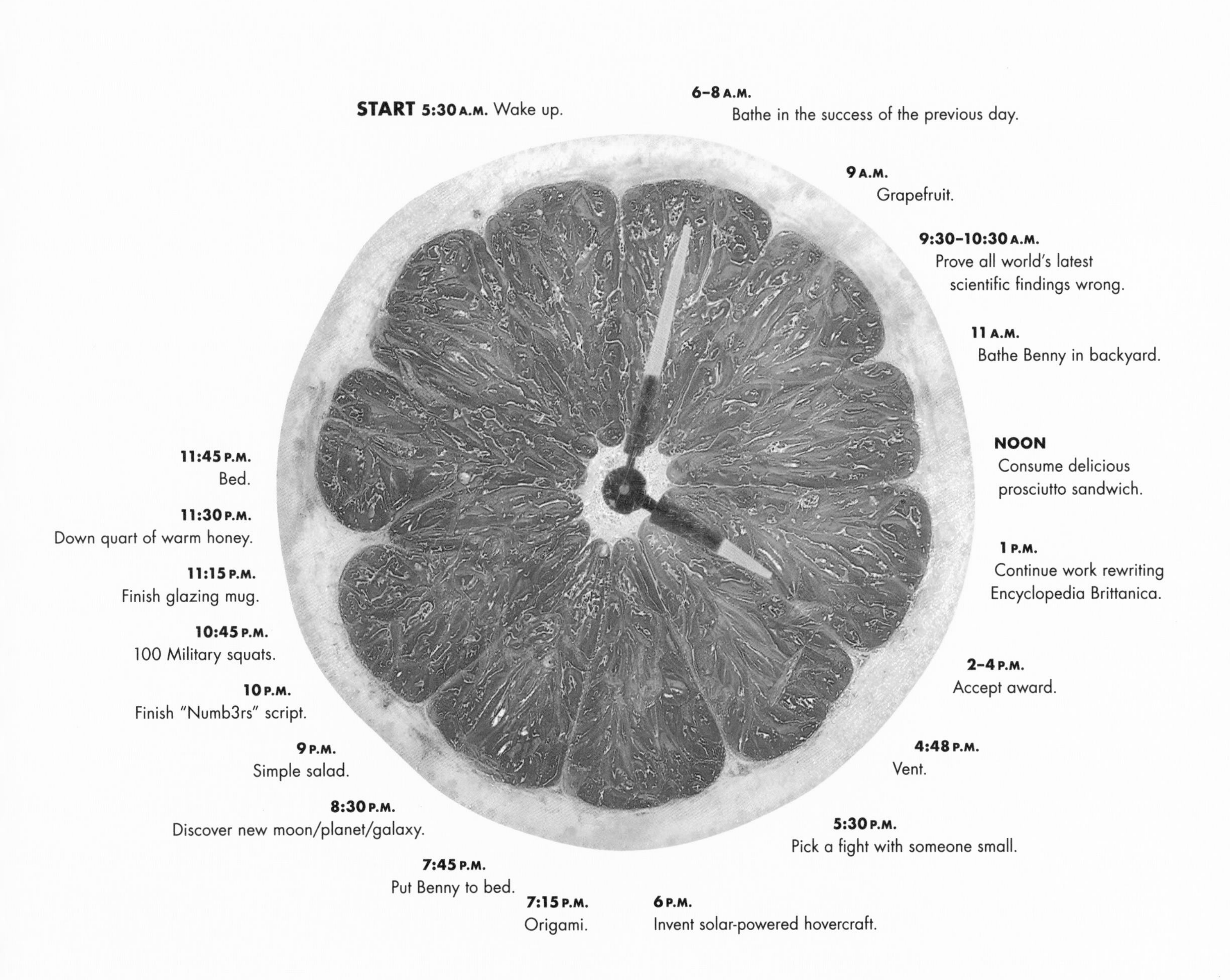

Now that you know about me, it is time to know about:

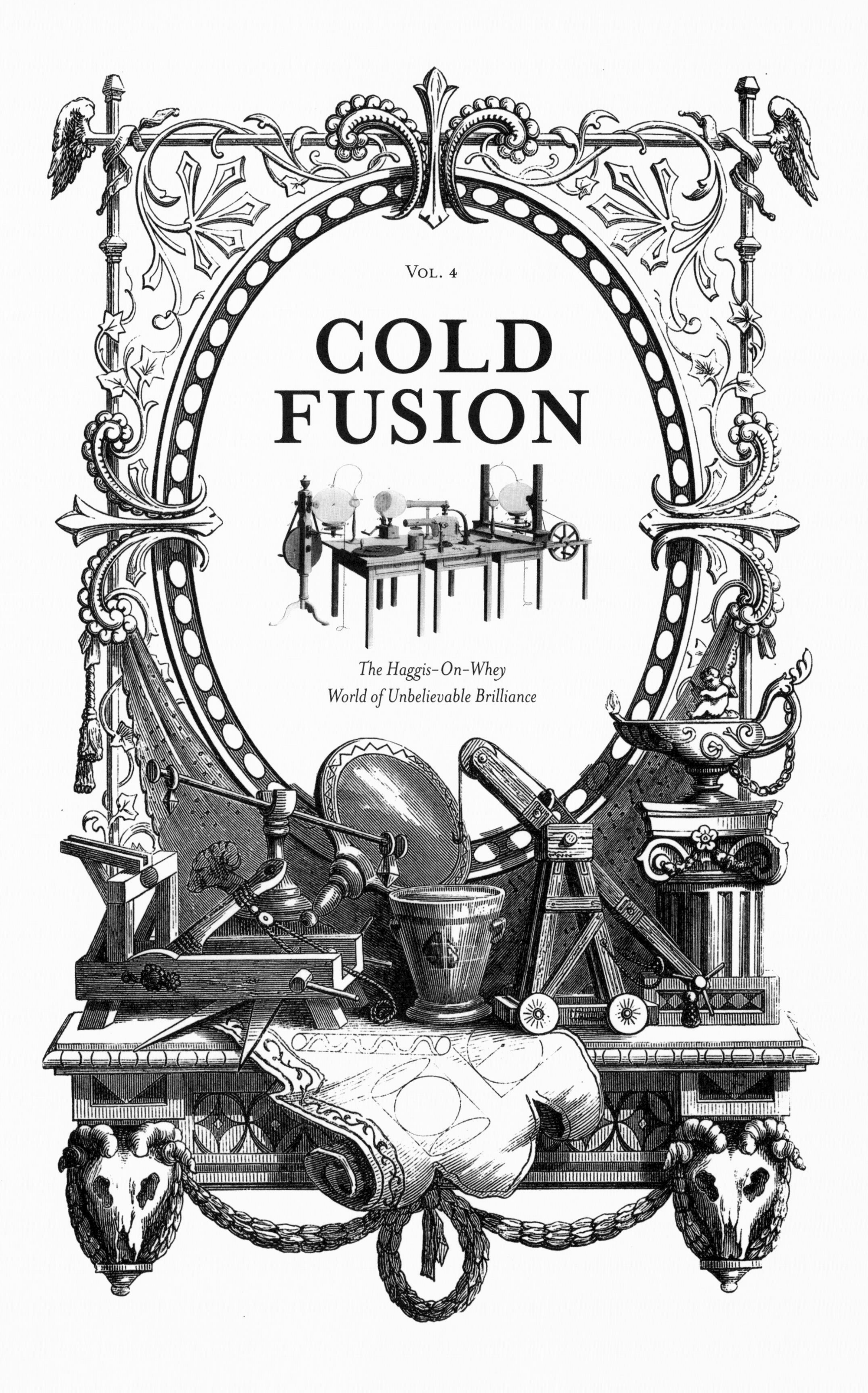

Vol. 4

COLD FUSION

The Haggis-On-Whey World of Unbelievable Brilliance

U.S.
POZZESE

THIS IS

The Most Important Book of Our Time, Your Time, and Their Time

Dear Unenlightened Reader,

Most of the air we breathe and the things in the world that make us happy come from cold fusion. That's why this book and probably the next two or three will be devoted entirely to the subject. Cold fusion is in fact so crucial to our very existence that only a few people know or care about it. Cold fusion affects reptiles, agriculture, the molecules we use to butter our toast, and television.

Imagine what you would do if your favorite swimming pool one day *evaporated*... into nothing. Now imagine if every pet you've ever owned *froze*... in the middle of space. And now imagine if instead of being *you*... you were somebody else entirely. Now stop and know this: cold fusion is more important, and more mysterious than all those things combined. And often louder.

People say we're at a dangerous time and that the cockroaches are the only ones that will survive, because of their hideous outer shell. Wrong. Cockroaches will survive forever because unlike today's leaders and yesterday's food stylists, they're giving in to cold fusion rather than trying to fight it. There's a lesson there.

There are three things every man and person is going to need to survive in the future: water, comfortable pants, and power. And if cold fusion's not one of those things, then we're in deep trouble. Maybe you're the type of person who tells guests to grab your mail on the way in. Maybe you're the type of person that likes watching someone's car backfire. Maybe you're the type who throws a cantaloupe in an egg fight. Whatever the case, someday soon, the time will come to pick sides. And that's the day cold fusion will come knocking. How are you going to answer the door?

I am so disappointed in you.

Signed,

Doris Haggis-On-Whey PhD, MS, DDS. BENNY

Dr. and Mr. Doris Haggis-On-Whey

COLD FUSION IN FIVE STEPS

STEP 1

Put on a white coat and white slacks. Wear a white nametag. Act like a professional.

STEP 2

Go into the bathroom. You will need the tub.

STEP 3

Gather a Dewar flask, an anode, and a cathode. Find some nice palladium or nickel. You'll need them in bulk—thin films or powder. Get a good amount of heavy water. Some people use plasma, but those people are imbeciles. Get some white shoes and put them on. Be ready to excite your palladium with electricity, magnetism, or a laser beam.

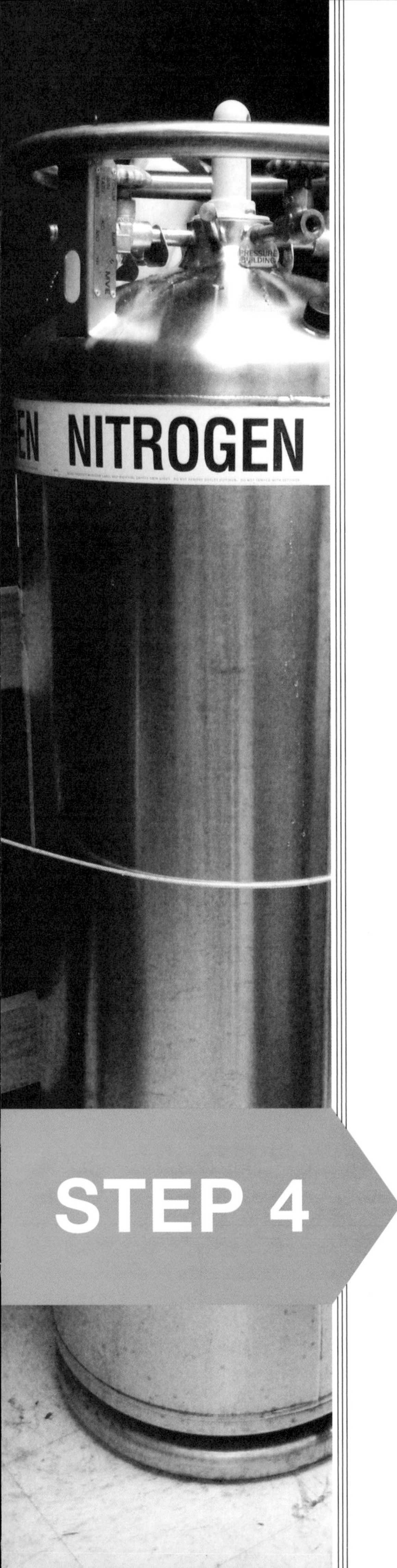

STEP 4

Now take your Dewar flask (a double-walled vacuum flask) and ensure that heat conduction will be minimal on the side and the bottom of the cell. Make sure that no more than five percent of the heat is lost in this experiment. The cell flask should then be submerged in a bath maintained at constant temperature to eliminate the effect of external heat sources. Now use an open cell, thus allowing the gaseous deuterium and oxygen resulting from the electrolysis reaction to leave the cell, along with some heat. It is of course necessary to replenish the cell with heavy water at regular intervals. Because your cell should be tall and narrow, the bubbling action of the gas will keep the electrolyte well mixed and of a uniform temperature. Special attention should be paid to the purity of the palladium cathode and electrolyte to prevent the build-up of material on its surface, especially after long periods of operation.

Your cell should also be instrumented with a thermistor to measure the temperature of the electrolyte, and an electrical heater to generate pulses of heat and calibrate the heat loss due to the gas outlet. After calibration, it will be possible to compute the heat generated by the reaction.

A constant current should be applied to the cell continuously for many weeks, and heavy water should be added as necessary. For most of the time, the power input to the cell should be equal to the power that leaves the cell within measuring accuracy, and the cell temperature should remain stable at around 30°C. You will see that at some point the temperature will rise suddenly to about 50°C without changes in the input power, for durations of two days or more. The generated power will be calculated to be about twenty times the input power during the power bursts. Eventually the power bursts in any one cell will no longer occur and the cell can be turned off.

Congratulations. You have therefore created a low-level nuclear fusion kind of thing.

STEP 5

Take pictures of the experiment—*using a good camera*. Write a good paper about your experiment. Double-space your paper and use 1" margins. Don't use some silly font; keep it serious-looking. Put a nice cover on it, something that will keep the paper from getting wrinkled or stained. Then send it to the serious scientific periodicals and your family.

FIVE PEOPLE WHO SCOFFED AT COLD FUSION AND WHAT HAPPENED TO THEM

TODD SHIPPE

Age 8

Offense: Accused cold fusion of being a great name for a WNBA team.

Result: Could no longer digest gluten. For the rest of his life he had to explain his condition to friends and acquaintances using the words "loose stool."

DOUG SHURE

Age 36

Offense: Said cold fusion gave him a bad taste in his mouth.

Result: Lived alone in a small apartment above a Subway and had his application for cat adoption denied.

INEZ MACHADO

Age 19

Offense: Asked why the scientists can't spend their time on something better like making Trident gum bigger or birds less fidgety.

Result: Lost four fingers to a family of possums when she stepped outside the tent to see the stars.

KIYOKA JOHANSEN

Age 11

Offense: Said it can't be that great because her dad's never heard of it.

Result: Cried whenever she sneezed and spent most of her 20s and 30s living in a grain mill.

FRED SANCHEZ

Age 68

Offense: Thought cold fusion was energy for jerks.

Result: Was gored by two children in a European elk costume.

THE HISTORY OF COLD FUSION

1926
German scientists F. Paneth and K. Peters report the transformation of hydrogen (H) into helium (He) by nuclear catalysis when H is absorbed by palladium at room temperature.

1927
Swedish scientist J. Tandberg reports the transformation of H into He in an electrolytic cell with palladium electrodes.

1932
After the discovery of deuterium, Tandberg continues his experiments with heavy water.

1934
P.I. Dee possibly discovers evidence for cold fusion.

1981
Dr. Mizuno observes charged particles from palladium deuterides, but attributes them to instrument error.

1984
Stanley Pons and Martin Fleischmann begin work on their basic experiments. The question they posed was "Would the nuclear reactions of deuterons confined in a lattice be faster (and different) from the fusion of deuterons in a plasma?"

1986
The term "cold fusion" is coined by Dr. Paul Palmer of Brigham Young University in an investigation of "geo-fusion."

1988
Pons and Fleischmann apply to the US Department of Energy for funding for a larger series of experiments.

1989
On March 23, chemists Stanley Pons and Martin Fleischmann announce that they have created fusion energy at room temperature in a simple test tube.

On April 10th, a team at Texas A&M University publish results of excess heat, and later that day a team at the Georgia Institute of Technology announce neutron production. Both teams ultimately withdraw their results due to lack of evidence.

April 12th—Pons receives a standing ovation at the meeting of the American Chemical Society.

April 28th—MIT Professors Ronald R. Parker and Ronald Ballinger plant an anti-cold fusion story in an interview published by the *Boston Herald*.

1989 CONTINUED
May 1st—The MIT News Office issues a denial of the *Herald's* reporting of Professor Parker's statements about Pons and Fleischmann's work as "scientific schlock" and "maybe fraud." Also, a series of failed cold fusion experiments are reported at the meeting of the American Physical Society.

In November, the Department of Energy releases its "Report on Cold Fusion Research." It concluded that useful sources of energy could not be obtained from cold fusion and that experiments to date lacked evidence to associate the reported excess heat with a nuclear reaction.

1949
Edward Joseph Mahoney is born on March 21 in New York City.

1964
Attends Island Trees High School in Levittown, Long Island, NY.

1968
Enrolls in the New York Police Academy to stay out of the Vietnam War. His father was a police officer.

EARLY 1970s
Money is kicked out of the Police Academy. He relocates to San Francisco and grows his hair long. He sells blue jeans and joins Big Brother and the Holding Company soon after Janis Joplin's demise.

1975
After spending years playing with different groups, Money becomes involved with Bay Area concert promoter Bill Graham's management company.

1976
With Billy Graham's help, Money signs with CBS Records.

1977
Money releases his first album, *Eddie Money*. The two most memorable hits of his career, "Baby Hold On" and "Two Tickets to Paradise," are on the album and they both enter the Top 40.

1978
Money releases his second album, *Life for the Taking*, featuring a pop/disco sound.

1980
Money releases his third album, *Playing for Keeps*.

1982
After a year-and-a-half hiatus from the public eye, Money releases his fourth album, *No Control*, in the summer. It hits #20 on the *Billboard* charts.

THE HISTORY OF EDDIE MONEY

1990
March—First International Conference on Cold Fusion held In Salt Lake City, Utah.

1991
In June, the National Cold Fusion Institute at the University of Utah closes and Dr. Pons leaves the University and moves to France where he sets up a lab with Fleischmann.

At the Como, Italy cold fusion conference in July, Pons and Fleischmann report that they brought an electrochemical solution to boiling point with 10–11 silver-palladium alloy electrodes.

1992
Eugene Mallove, the former chief science writer at the MIT News Office, presents a petition to Congress urging Federal funding of cold fusion research.

While derided in the US, cold fusion research is embraced in Japan. The Japanese Ministry of International Trade and industry finance cold fusion research at the level of $24 million.

On January 27th, Dr. Akito Takahashi reports the production of ~70 watts of excess heat over a one-month period from a 1mm thick x 35mm x 35mm palladium plate. The total excess heat calculated to 200 megajoules, which is orders of magnitude beyond what any chemical reaction can produce.

On August 15th, Dr. Edmun Storms of the Los Alamos National Laboratory announces that he has replicated Takahashi's experiment.

1993
Les Case mixes carbon, palladium, and deuterium and generates 5 ppm helium and excess heat suggesting the presence of a nuclear reaction.

On May 5th, the Energy Subcommittee of the US House Committee on Science, Space, and Technology agrees to allow the DOE to fund cold fusion research, marking a turning point from Congressional inactivity.

At the Fourth International Conference on Cold Fusion, observations such as the transmutation of heavy elements at low energy lead the field in new directions.

1994
In August, the Electric Power Research Institute (EPRI), which had been financing cold fusion research at SRI International, reports that nuclear reactions have been detected at levels forty orders of magnitude greater than predicted by conventional nuclear theory.

1996
In February, a NASA Technical Memorandum concludes that "replication of experiments claiming to demonstrate excess heat production in light water-Ni-K2CO3 electrolytic cells was found to produce an apparent excess heat of 11 W maximum, for 60 W electrical power into the cell."

1997
In *The Saint*, Val Kilmer plays master of disguise Simon Templar, hired by the Russians to steal a formula for cold fusion from a distractingly beautiful physicist (Elisabeth Shue).

2004
In December, the DOE releases a second cold fusion report that does not recommend Congressional funding for cold fusion research, but identifies areas of research that could resolve some of the conflicts and controversies within the field.

1983
Money's lowest-charting album, *Where's The Party?*, is released.

1986
Can't Hold Back is released and eventually goes platinum.

1988
Nothing To Lose is released.

1989
Money takes a break to concentrate on his family. He gets married and has five children.

1991
Right Here, Money's eighth studio album, is released.

1992
Money releases an acoustic album entitled *Unplug It In*. Small club dates of acoustic sets follow its release.

1995
Love & Money is released and considered a comeback for Eddie Money.

2001
Money plays himself in the classic movie *Joe Dirt*. He is overlooked by the Academy.

2004
"Two Tickets to Paradise" is featured in the game *Grand Theft Auto: San Andreas*. On February 24th, Money performs the national anthem at a game between the Detroit Pistons and the Minnesota Timberwolves.

2007
Money's most recent album, *Wanna Go Back*, is an album of cover songs from his high school days with his first band, The Grapes of Wrath.

THE HISTORY OF TIMELINES

600 B.C.
First timeline invented.

350 B.C.
Second timeline made.

3 B.C.
Dashes added.

620 A.D.
Italics used.

1175 A.D.
Font changed.

1550 A.D.
Bold type incorporated.

1555 A.D.
Colons introduced.

1981 A.D.
Timelines banned from all ha-ha books.

SOME THINGS MADE IMMEDIATELY BETTER WITH COLD FUSION

Reggae Music

Covens

Back Taxes

Interracial Dating

German Guilt

The West Indies

Mannequin Fights

Fruit Bats

Photosynthesis

Shepherd's Pie

Comebacks

Instrumentals

Flag Day

Woven Wheats

Suit Jackets or BLAZERS

AMOUNTS OF MONEY DIFFERENT GROUPS CAN MAKE USING COLD FUSION

Gradually, people will see that money can be made using cold fusion. Companies will begin to make something useful that they can sell. Once that happens, development will be rapid.

PERSON	AMOUNT
Catholics *(regular)*	$14,000
Catholics *(Irish)*	$9,000
Buddhists *(regular)*	$12,500
Buddhists *(Irish)*	$9,000
Buddhists *(Californian)*	$1,000
Agnostics	$4,400
Tree farmers	$70,000
Trees	$98,500
Sea Captains	$32,000
Weekend boaters	$45,000
Politicians *(regular)*	$6,000
Politicians *(female)*	$6,500
Politicians *(obese)*	$6,600
Diggers of wells	$41,000
Diggers of ditches	$41,000
Diggers of Eddie Money	$5,300
People named Eddie Money	$32,000
People named Kyle MacLachlan	$22,200
People shaped like seals	$66,000
People wronged by seals	$5,400

ART INSPIRED BY COLD FUSION

Ballerina II.

Miró. *Ballerina II.* 1925. The blues in this work are obvious homages to the tritium byproducts in a heavy metal/heavy water synthesis. An observant scientist can note how Miró has painted an exact depiction of a helium electron spiraling away at the apex of the charged reaction. Miró always noted that he *owed* much inspiration to the scientific world, but in this work one wonders if the word *plagiarism* might be more appropriate.

Warhol. *200 Campbell's Soup Cans.* 1962. While many have interpreted the repetition as a comment on mass-production and consumerism, reading Warhol's science-filled diaries we see that it is instead a reference to the anomalous transmutations and isotope shifts that are at the heart of the fourth stage of cold fusion.

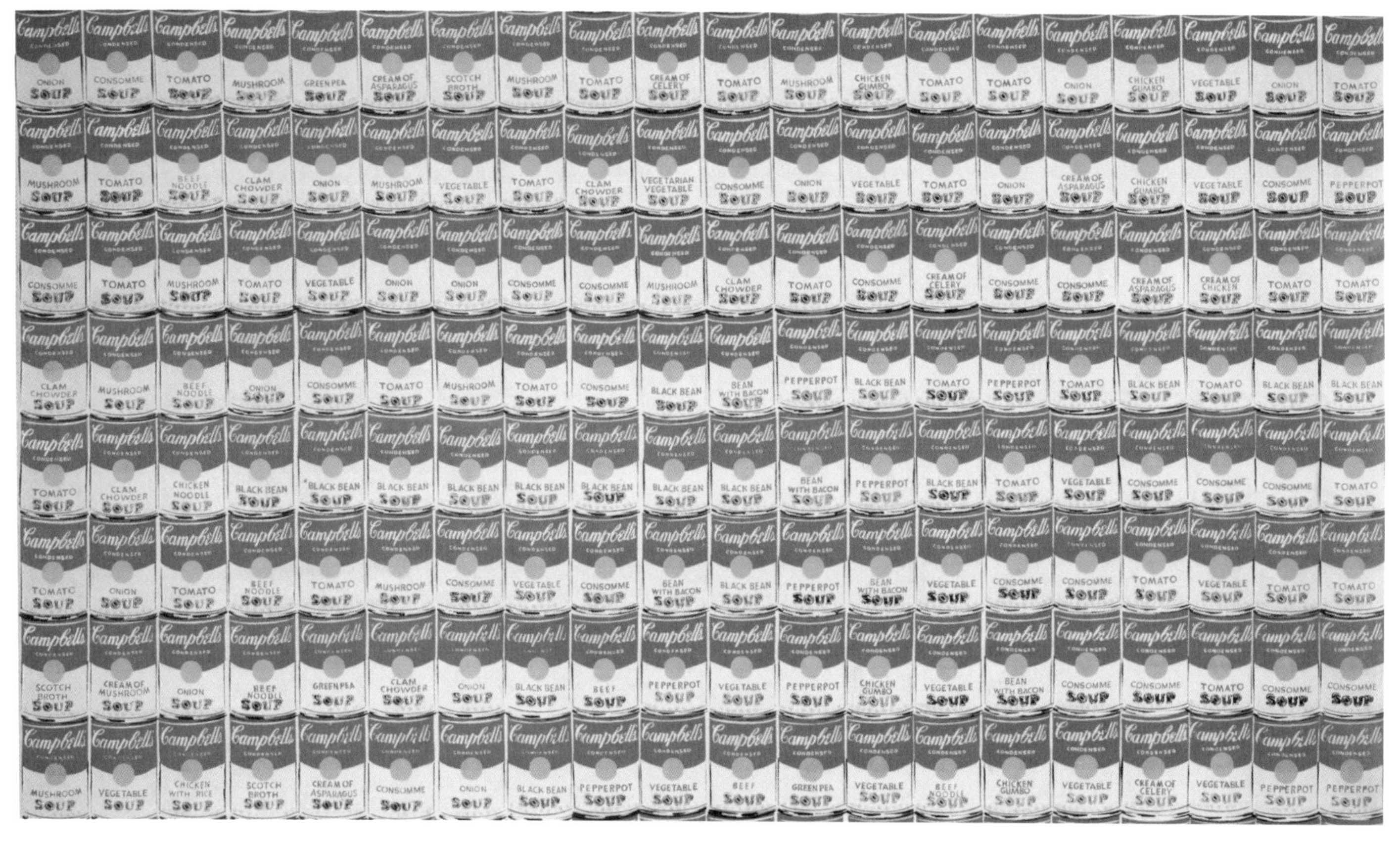

200 Campbell's Soup Cans.

Drowning Girl.

Lichtenstein. *Drowning Girl.* 1963. In his seminal work, *Drowning Girl,* the woman in the painting cries, "I don't care! I'd rather sink—than call Brad for help!" Here, of course, the figure of 'Brad' is a clear allusion to Frederick Bradislav, the Russian who first proposed the concept of nuclear fusion at near-room temperatures. And of course the woman is drowning in water much as hydrogen atoms would fuse together in heavy water. Finally, as if it were not obvious enough, Lichtenstein barely hides an excessive energy differential equation in the water.

Bosch. *The Garden of Earthly Delights.* 1503–4. While this painting is largely about the dichotomy of good and evil, the saved and the doomed, it is also about a proposed form of safe and efficient nuclear energy six hundred years before its time. If one can imagine, as Bosch himself did, the man with the legs of a robin as piezoelectric substrate and the bird-faced, serpent-bodied figure eating a human leg on top of a wooden pedestal as two deuterium atoms hot-fusing together, then the painting really starts to resemble less a painting and more a fairly standard—banal, even—chemical reaction.

The Garden of Earthly Delights.

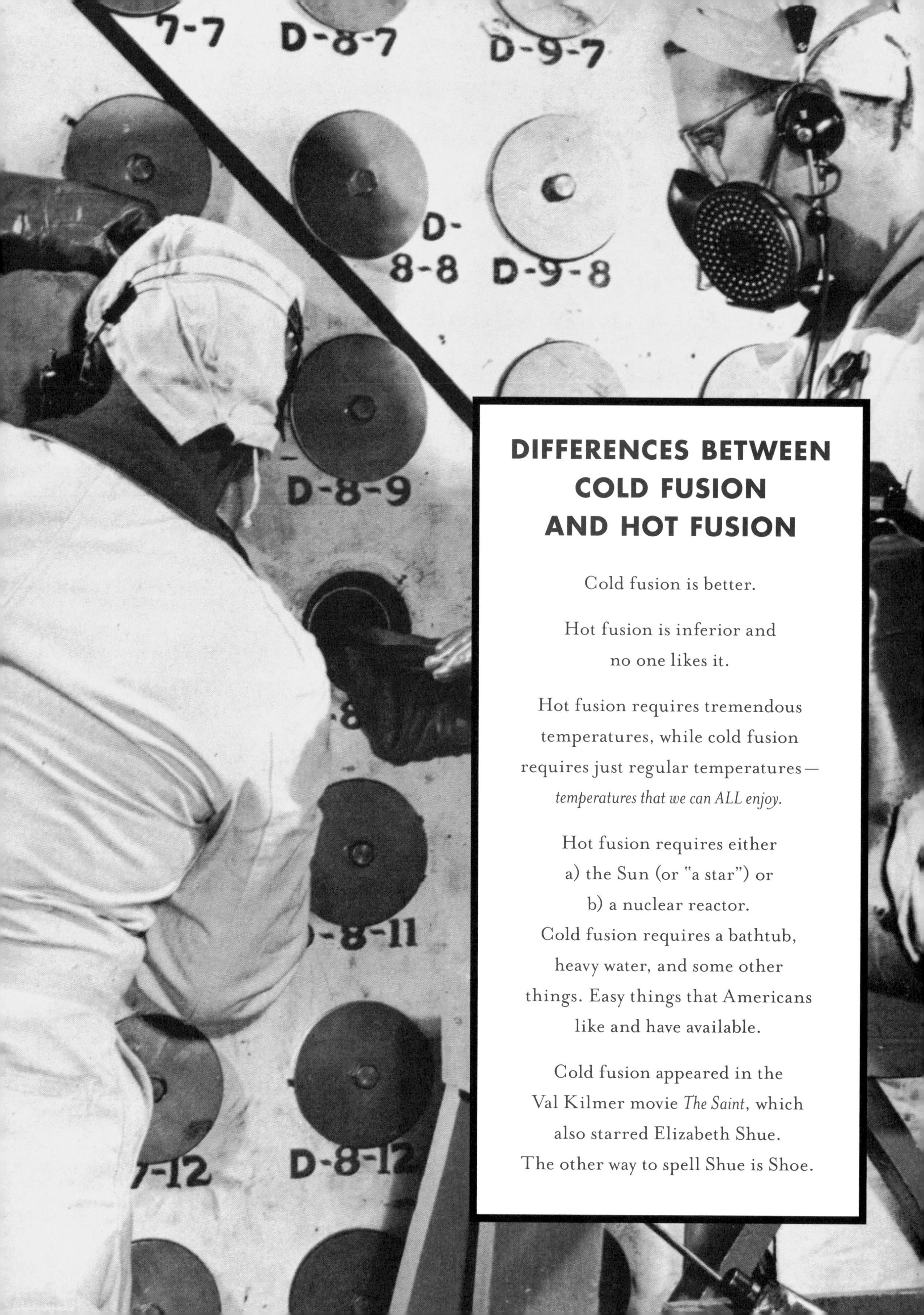

DIFFERENCES BETWEEN COLD FUSION AND HOT FUSION

Cold fusion is better.

Hot fusion is inferior and no one likes it.

Hot fusion requires tremendous temperatures, while cold fusion requires just regular temperatures—*temperatures that we can ALL enjoy.*

Hot fusion requires either
a) the Sun (or "a star") or
b) a nuclear reactor.
Cold fusion requires a bathtub, heavy water, and some other things. Easy things that Americans like and have available.

Cold fusion appeared in the Val Kilmer movie *The Saint*, which also starred Elizabeth Shue. The other way to spell Shue is Shoe.

100 PEOPLE WHO WILL NEVER BE GOOD AT COLD FUSION

PERSON	REASON
MATTHEW PERRY	Actor
CHRISTO	Artist
GEN. DWIGHT D. EISENHOWER	Dead
TIM CONWAY	Comedian
HELEN MIRREN	Bad Oscar speech
GEN. PERVEZ MUSHARRAF	Name "Pervez"
JULIA CHILD	Dead
MAXIMILLIAN, *the robot from* The Black Hole	Too butch
RONALD REAGAN	Too butch
THE VON TRAPP FAMILY	Too Austrian
ALL AUSTRIANS	All the children wear hats
ALL FINNS	All the women wear corduroy
ALL SWEDES	All the men wear glasses and have weird little knees
MATTHEW PERRY	Did we already say Matthew Perry?
STEVE PERRY	Too good
STEVE PERRY'S REPLACEMENTS	Not good enough
ALL SAILORS	Too briny
ANNE BAXTER	So creepy in *All About Eve*
MIKE HUCKABEE	Who is Mike Huckabee?
RESIDENTS OF LANCASTER, PA	Too horsey
HELEN MIRREN	Definitely not her
HELEN HAYES	Reminds us of Helen Mirren
HELEN HUNT	No alliterative Helens at all
GRADUATES OF BRYN MAWR	Too plaid
ADVOCATES OF EVAN BAYH	Too Gary

THE BOX IN HIS HEAD THAT GALILEO PUT COLD FUSION INTO

There came a day, a summer day, in the middle of August, in one of the many years in which Galileo Galilei lived, that he, the father of modern science, had to either announce his monumental discovery of the nuclear power known as cold fusion, or bottle it up in the recesses of his mind. On that day, for reasons never revealed until this publication, he corked it.

The story in 568 words:

Galileo was, one day while he lived, attending a celebration held by people with money to congratulate people like himself for being smart. Naturally he was wearing green. Galileo was laughing because he had just heard a lewd joke delivered by Francesco Sizi.

Galileo with Todd.

Suddenly Sir Francis Drake pulled Galileo toward him and slurred, "Each man has six great ideas," and then staggered away.

Galileo put his cracker search on pause—he had been looking for crackers—to contemplate this rare nugget of wisdom from the English pirate. He thought back on his monumental accomplishments—the telescope, the orientation of the modern solar system, kinematics, the thermometer, and his reinvention of the compass. "By god," said the red-cheeked Galileo, "I've only got one great idea left! This certainly is my feet of clay!"

This rhetorical utterance, though strange to us, was a common phrase referring to the pinch of a situation Sir Francis had put Galileo into. His head was awhirl. He thought of all his grade-A thoughts, and soon decided that there were two that soared above the rest. One of these two soaring notions was cold fusion. It was an elegant process, certainly, and would change the world utterly, but then again, it had such a silly name, and where would he find palladium? And besides, wasn't that the name of a nightclub in Chicago?

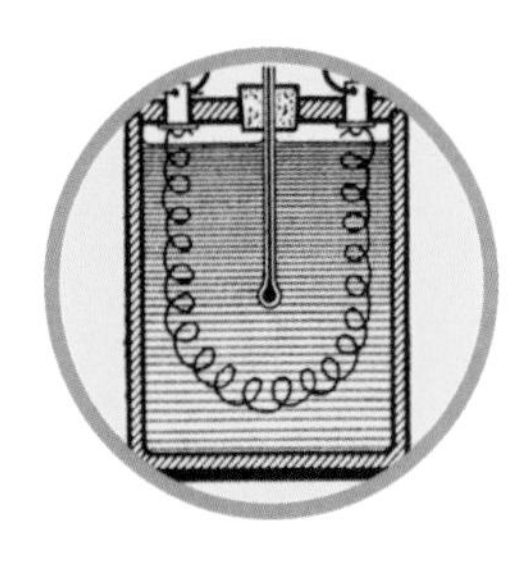

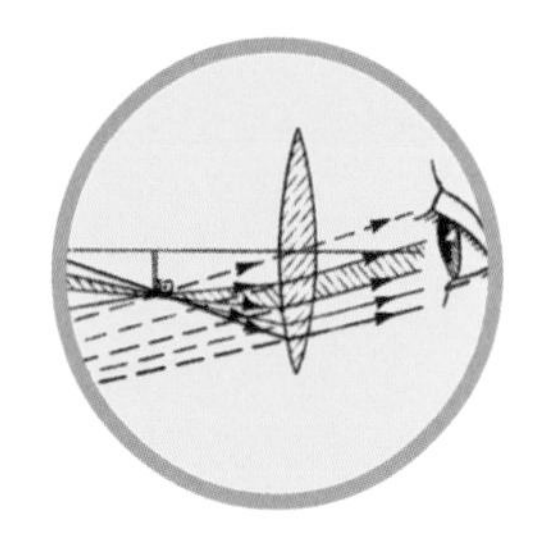

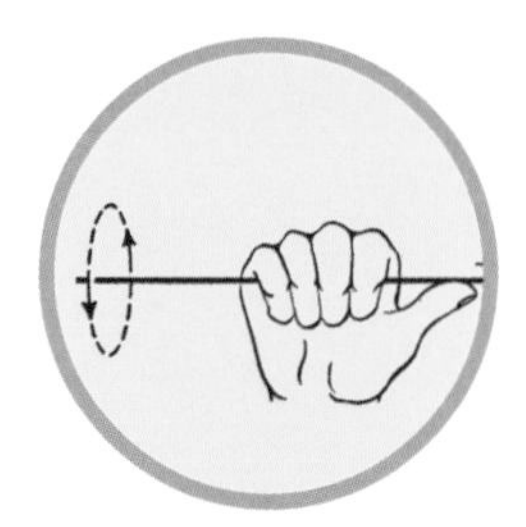

And so, with a deep sigh, Galileo did what no scientist outside of Max Planck would ever dare: he took a theory and put it away, deep back inside a small, wood-carved trunk of ideas that he locked in the deepest chasm of his round Italian head, never to be heard of again.

And with that, Galileo set off to rejoin the party at hand. But before he did, he sat down in the drawing room—his drawing room was just off to the side of the party, which was in his house—and he did the first sketches for his last great idea, which was of course pants that become more taut around the waist when one pulls two ends of the same string. *Draw-string pants*, as we now know them, were born on that day. The party was during the daytime. It was a daytime party.

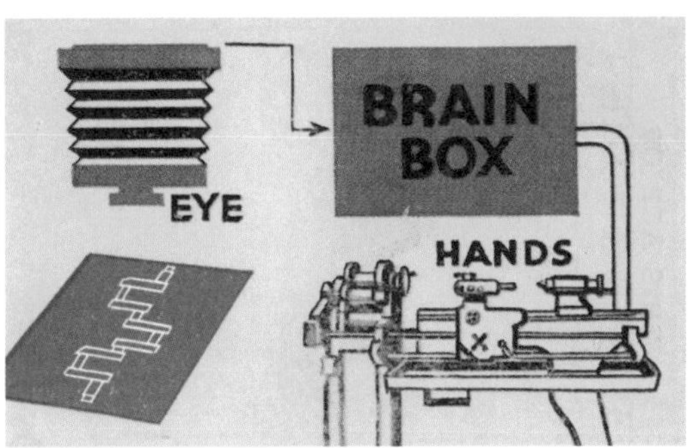

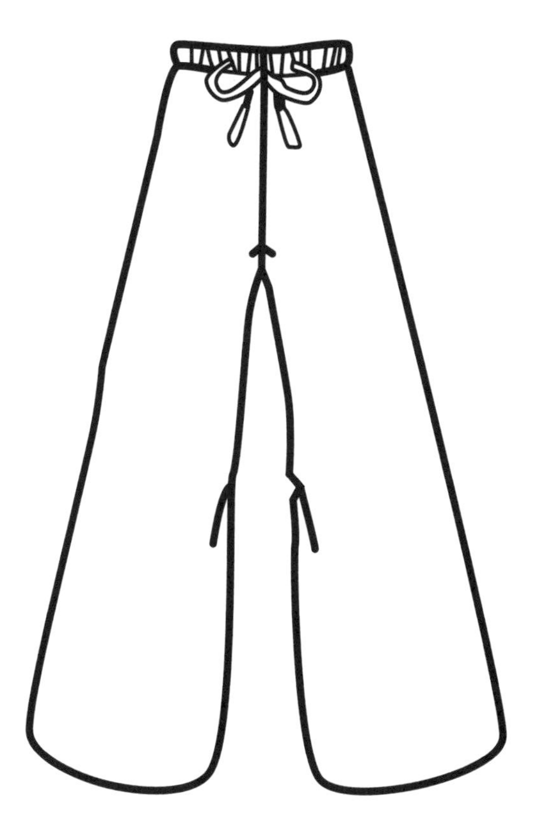

PARTS OF THE HOUSE BEST SUITED TO LOW-ENERGY NUCLEAR TRANSFORMATIONS

CHILDREN'S BEDROOM Most modern children's bedrooms are equipped with palladium delivery systems and lasers, so this is a natural "first choice" for experiments. Night-time, while the occupants sleep, is the best time—though success might evoke a low-level chartreuse or tangerine light which might wake light sleepers.

DINING ROOM Scientists at Texas A&M, including the man called John Bockris, famously used their dining room (and those of their neighbors) for their work. Dining rooms make sense because the atmosphere is pleasant and they are usually situated near the front door, allowing for quick exits if radiation threatens life or linen.

KITCHEN This room is not recommended.

GARAGE This large venue should be a perfect environment to store capsules of coated carbon pellets, deuterium gas reserves, and the other essential ingredients you have somehow procured funding for. Garages are also cold and dreary.

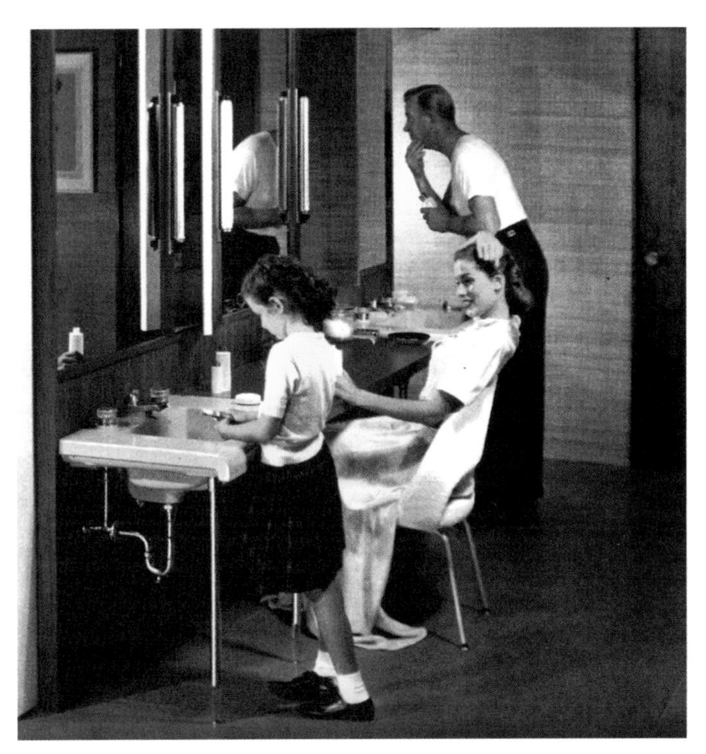

BATHROOM This is a good room, as you might need a faucet into which you can wash your mistakes.

IN-LAW APARTMENT While not ideal, with the addition of some faulty insulation, bolts on the windows, and towels stuffed under the door, you can nearly replicate the environment in which Kafka did most of his writing. On the downside, it was in there that he also invented the pumpkin peeler, which is actually worse than it sounds.

LIVING ROOM The typical American living room is where the largest windows in the house are usually placed. This would open the slight possibility that the accidental fireball you create will exit through these large windows—instead of swallowing you and the house whole.

LABORATORY PRANKS

1. Take one of the modified bison livers from the incubator and throw it into the garbage. Next, generously apply Bullseye BBQ sauce to your lips and shirt. When your partner asks you where the missing liver is, look confused and say, "If that was modified bison liver, then where the heck is my lunch?"

2. While your lab partner is taking a nap, roll him up in a periodic table poster, fasten a belt of polyethylene tubing around him, and leave him in the courtyard with a note attached to his shirt saying "Hypothesis Failed."

3. Pretend to mix sodium hydroxide and hydrochloric acid, when in reality it's really just Kool-Aid. Later that day, pretend to have amnesia. Then pick up the mixed beaker, ask your lab partner what it is and before he/she can answer put it up to your mouth and say, "There's only one way to find out for sure," and drink it down in one gulp. Fake cardiac arrest. Right before he/she hits you with a defibrillator, yell "Psych!"

4. Spend a week rounding up all numbers in the lab's daily calculations. Then when your lab partners are driven crazy because the numbers don't add up, tell them what you did and then say to them, "What can I say? I'm an optimist!"

5. Ask your assistant for a left-handed cathode ray.

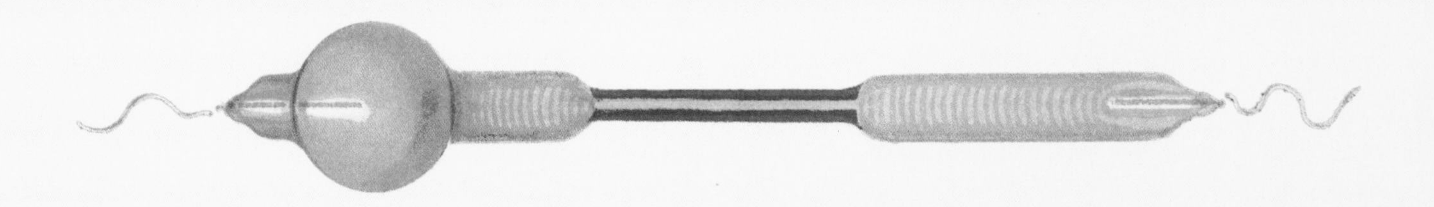

6. SPEND FIVE MONTHS meticulously teaching a mouse how to pick open a briefcase's simple combination lock. Then, for Christmas, buy your lab partner a briefcase with a simple combination lock. Next, over the next few months, convince him of the safety of this lock until he finally entrusts all his most valuable possessions to it. Then, have the mouse rob him blind during a lunch break. When your partner looks for his stolen possessions, shrug your shoulders and say, "Told you we should have used rats!"

7. MAKE A VOICE RECORDING of a high-pitched voice saying "Why?" and put it into a remote-controlled voice box. Put this near the animal cages. Then, every time your lab partner pulls a rat out to experiment on, activate the recording. They'll believe it's the rats talking. Or maybe they won't. This prank isn't as good as some of the others.

8. COME INTO WORK EARLY one day after shaving your eyebrows. Then rub dirt and oil all over your face and have your back turned to everyone when they start to file in. At the precise moment when they're all near you, drop a few blocks of dry ice into a container of water and soap. When the harmless explosion engulfs you, yell "Arrrgh! The pain! I'm blind!" and collapse under the eyewash station. Hold for laughs. Or lunch.

WILL HISTORY REMEMBER YOU OR NOT? HOW TO AVOID BECOMING ANOTHER JOHANNES KEPLER

If you are not stupid, someday you will make a world-changing scientific discovery that will alter the landscape of all human endeavor. But even when you do so—if you do not fail, which you very well might—discovery is only half of the battle. After publishing your work and being showered with accolades, now the real work to begin: cementing your legacy. There are hordes of scientists in the history who discovered something of great importance but who few have heard of. Here's a name: James Clark Maxwell. Sounds like an English actor playing King Lear on PBS, in 1981. Actually, he invented the magnetic field. Here's another name: Ernest Rutherford. He was not a mid-level pig farmer from the South but the very father of nuclear physics.

On the other side there's Rudolph Polio, William Phonograph, Franklin T. Lamp and of course Thomas A. Vacuum. These are men who understood legacy. Which leads us to Johannes Kepler. Kepler trumped the accomplishments of contemporaries Copernicus and Galileo but nobody has named middle schools after him, and he has never been dramatized in independent film. Kepler invented the color white, the little symbol that indicates radioactivity, and of course the volcano. But the only thing he's known for now is wearing a jerkin on the 10 euro which, of course, is the world's only store-accepted form of play money. Absorb and ponder.

Thomas A. Vacuum

SO HOW CAN YOU AVOID PULLING A KEPLER?

Step 1: Name the device or disease—and every device or disease you create or discover—after yourself. Check your encyclopedia; there is no subtlety in the pages of history.

Step 2: Anger the church. (Not too difficult; they *hated* Thomas A. Vacuum.) In 400 years, when they pardon you, you will be remembered anew.

Step 3: Start taking a day off in your own honor on the same day every year. Hope others will eventually follow suit. The day will be named after you, as it was with Casimir Pulaski and Reginald G. Independence.

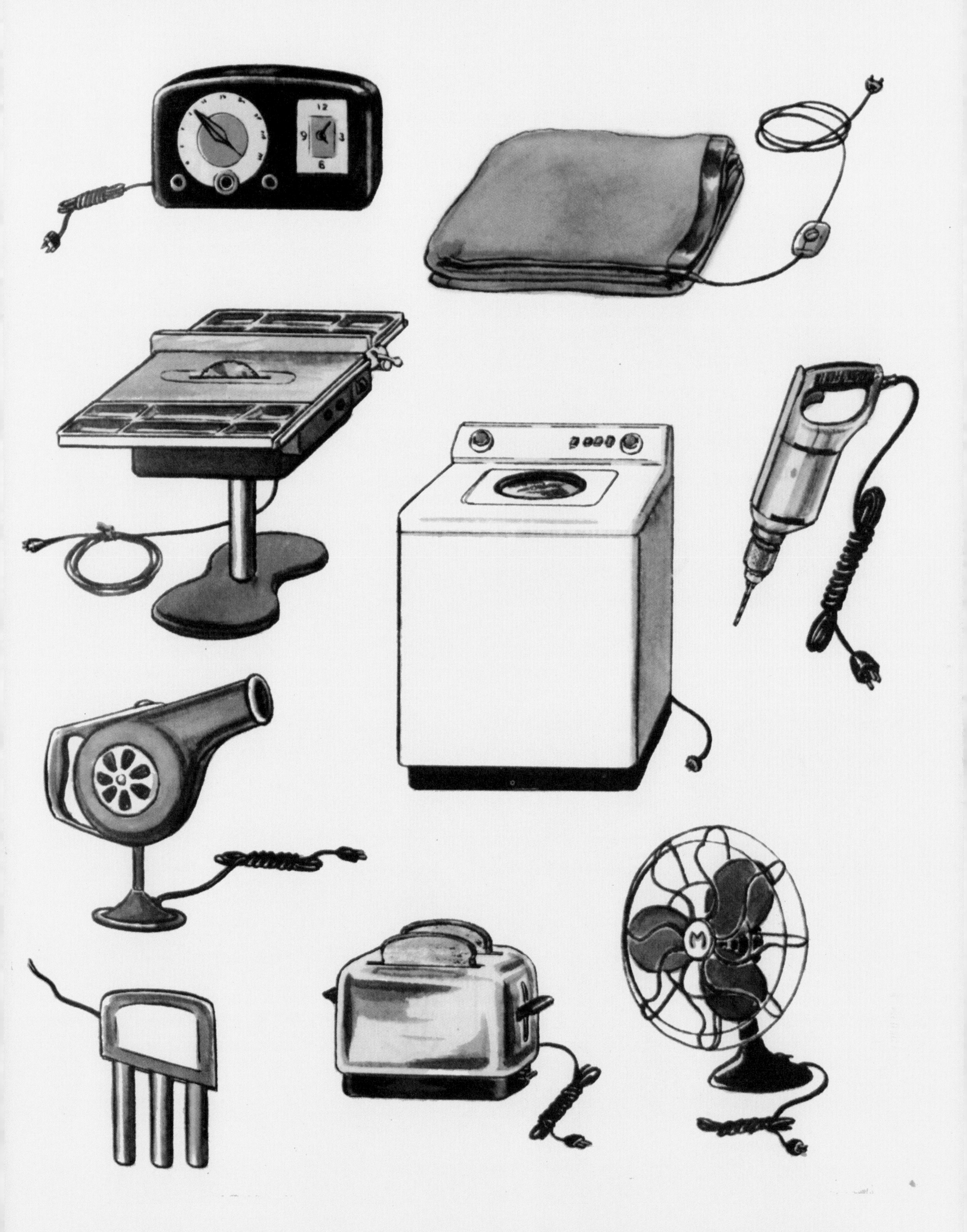

COMMON HOUSEHOLD DEVICES NAMED AFTER THEIR INVENTOR

POSSIBLE SIDE EFFECTS OF ROOM-TEMPERATURE NUCLEAR REACTIONS

Facial hair planning errors.

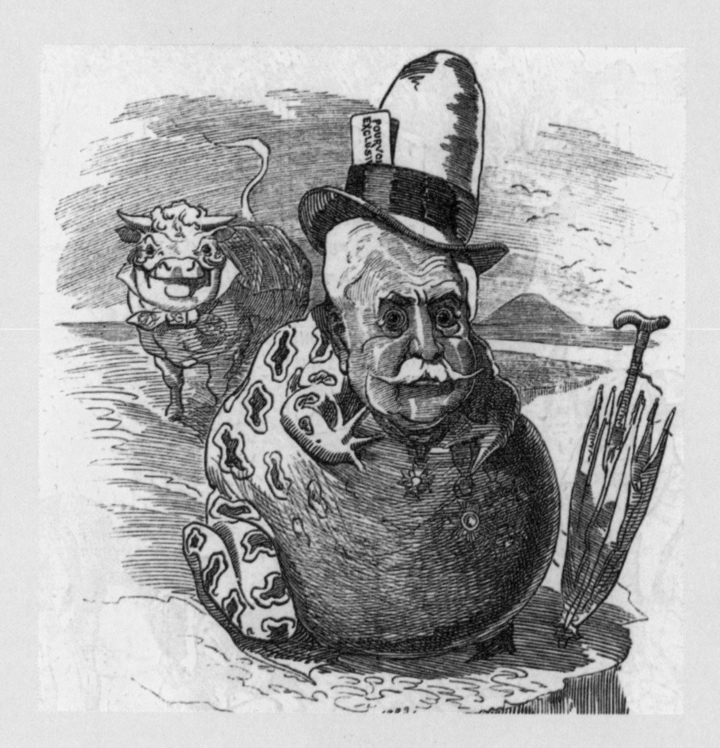

Frogmen.

Women in etchings are jailed for headaches.

Arctic regions no longer narrated.

Kevin Spacey stars in another movie set in Nova Scotia. Smash hit!

Sleeves of Emily Dickinson grow more ornate and helpful to the writing of verse.

Pumpkins become black and need to be measured.

Letters illuminated by flashing bulbs grow dimmer; people enticed by letters illuminated by flashing bulbs find life less interesting.

THE NIGHT CONDENSED MATTER NUCLEAR SCIENCE ENTERED THE TOWN OF DRAX, VA

It was a cold quiet night, devoid of any noise or sunlight. All the children had been laid down on their wood planks, kissed on the cheek and pushed into their cupboards for the night. The dogs had been put outside along with the milk bottles and the sheep had begun tallying their day's take at the nearby OTB.

In all ways, from the shuttered windows to streets soaked in butter, it was as typical an evening as there was to be had in Drax. Then condensed matter nuclear science appeared.

Like all good science, it entered silently and invisibly. The mayor, suffering from a severe bout of insomnia or rickets, walked down the main street of town and tried to recount the last time he dreamt. Condensed matter nuclear science passed right by and when it did, the mayor felt a feeling he'd only felt once before. Quickly he ran off to his home to hide under his sink, worried his ex-girlfriend had returned to talk, again and at length, about why she had broken up with him.

The village alchemist was up, also unable to sleep. As was often the case, he was sitting in front of his lab, fretting over having an occupation that no longer existed. Condensed matter nuclear science passed by his window. The alchemist's pupils widened. He washed his hands in milk and began to chant like the schoolgirl he'd once been.

The swingsets rustled, and the town dirigible fluttered as condensed matter nuclear science moved further into town. Old Mrs. Jacoby, shrouded in a shroud, felt it pass her house. As it did she smiled. Old Mrs. Jacoby smiled because she knew no better, and was crazy, and was as smart as a table.

Condensed matter nuclear science continued to travel in ambiguous form throughout the town, altering all in its path. From the mechanical pencils, to non-mechanized ones, everything in the town felt the mark of this mysterious entity, without been able to see these changes at all. For years after this day, the townspeople would gather together and speak of what had transpired that night. Which was of course nothing, but their stories sounded compelling when delivered in their adorable Slavic accent.

COLD FUSION IN EPIC VERSE

Plainly speaking, there has been far too much epic poetry about fission and fusion, and most of it is terrible. It was written by men in the 17th century, wearing elf shoes and eyeliner. But there are a few examples of cold-fusion verse that are worth mentioning. Below find one such example. It is bad.

Sun's Everto Sought Sparrow

by Lord Byron*

O Sparrow! O Sparrow!
Thy fliest now with such delight and pleasant aviary mirth
That thou lighten the morn from and to reaches farthest of this earth
Oh! Sparrow, wise sparrow, fortitude was not yours to take
Too oblivious were you, to see the gods casually sharpening your stake,
Sparrow, blessed sparrow,
wanderest you through woods quite forlorn
Say! Make thy bread fresh lest thy wings be torn
Hey! I thought you beautiful but instead you said pretty
Knowest you ever the love that so arrogantly springs forth, witty
Chamomile is nice, but your spiced tea cannot be matched
From wherefor springs the soul's wicked black hole that cannot be patched?

Sparrow, oh winged sparrow,
Cold fusion's first outing, really will be quite a day
Durst though ask, 'why is that?' then leadest you, the fool's parade
Sparrow, blessed sparrow,
So small not undst like the straw
Williest thou wed thy beak with my paw?
Oh Sparrow or sparrow, seeist yourself as I see you see me?
If thou doth not protest let us pull anchor and together be free
I think cold fusion will solve the planet's energy problems
and will be quite safe and will not cause black holes
Thank you for your attention.

---

* Lord Byron was a very short man, no more than four feet high—about two meters—who wrote things about people and certain days. He never married, and was arguably a bachelor. Nothing he wrote survives today, nor do his friends, of which he had none. He was a bowler and a liar.

TROUBLESHOOTING WITH COLD FUSION

Your experiments to produce nuclear fusion at room temperatures haven't worked. Why? You may be a fool. You may very well be a fool. But before concluding you are a fool, check these questions and reminders.

Were you wearing white pants? A white nametag? We told you about this on page 7.

Did you use an open cell, thus allowing the gaseous deuterium and oxygen resulting from the electrolysis reaction to leave the cell, along with some heat? If you didn't do that you shouldn't be doing this kind of experiment at all.

Did you pay special attention to the purity of the palladium cathode and electrolyte? This prevents the build-up of material on its surface, especially after long periods of operation. No one should have to tell you that. This is getting pathetic.

Did you instrument your cell with a thermistor to measure the temperature of the electrolyte? And an electrical heater to generate pulses of heat and calibrate the heat loss due to the gas outlet? Any idiot would have done that.

Are your calorimeters sophisticated? Your calorimeters should be sophisticated.

Do you teach at Brigham Young University? You should teach somewhere else now.

DO YOU ALSO
USE A
MASS SPECTROMETER
TO CONFIRM
THAT THE SAMPLE
IS TRITIUM?

NO.

THE BEST KINDS OF FOLDERS AND BINDERS IN WHICH TO PUT YOUR COLD FUSION PLANS

1″ Avery Durable View Binder with EZ-Turn Rings, White

Durable construction is ideal for frequent reference. **$4.49**

1″ Avery Heavy-Duty View Binder with One Touch EZD Rings, White

One Touch rings open and close with the press of a finger. **$6.23**

1″ Avery Heavy Duty Nonstick View Binder, White

Nonstick material ensures no ink or toner transfer from printed materials to clear overlay, pockets, or panels. **$8.29**

1-1/2″ Staples Plastic Frosted Binders with Round Rings, Blue

Durable, heavy-gauge polypropylene material resists dirt and moisture. **$5.79**

1″ Avery Poly Binders with Round Rings, Teal

Lightweight and flexible. **$2.15**

1″ Avery Flexi-View Binder, Grey

Avery Flexible Presentation Binder has a stylish graphic border on the cover that helps your work look professional and neat. **$3.89**

Wilsonjones Professional View-Tab 3-Ring Binder with Subject Dividers

Soft, rounded spine for comfortable grip and professional look. **$7.35**

1" Samsill Two-Tone D-Ring Binder with Labelholder, Purple/Turquoise

Patented Flexaround construction prevents top-loading sheet protectors and oversize indexes from extending outside binder. **$10.99**

Case-It Dual-100 Dual Binder 2-in-1 Zipper Binder, Red

Features patent-pending Velcro dividers to secure contents and a removable zippered pencil pouch. **$19.99**

Mead Five Star 2-Pocket Laminate Portfolio

Conversion tables and other useful information inside front and back cover. **$1.69**

Staples 2-Pocket Portfolio with 3-Prong Fasteners

Double-ply reinforced pockets and folder top provide extra strength. **$.49**

Oxford 2-Ply File Folder with Straight Cut

Perfect for color-coding files in active professional offices. **$.35**

PAID TESTIMONIALS FROM ENTHUSIASTS AND SURVEYORS

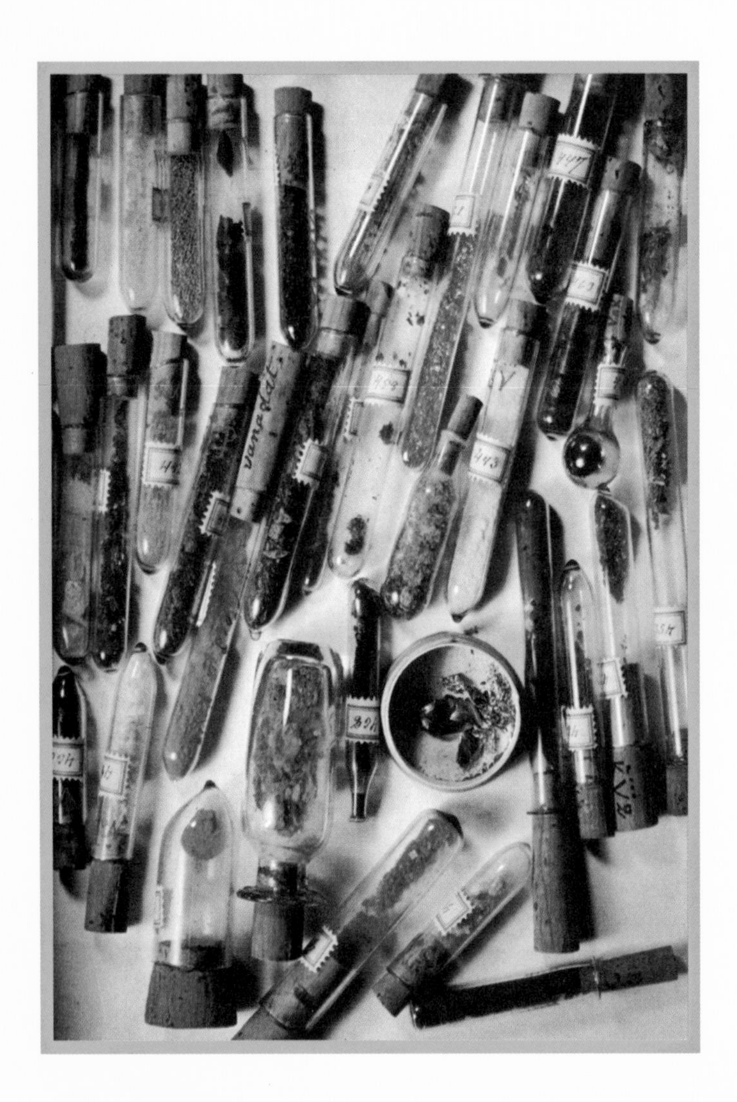

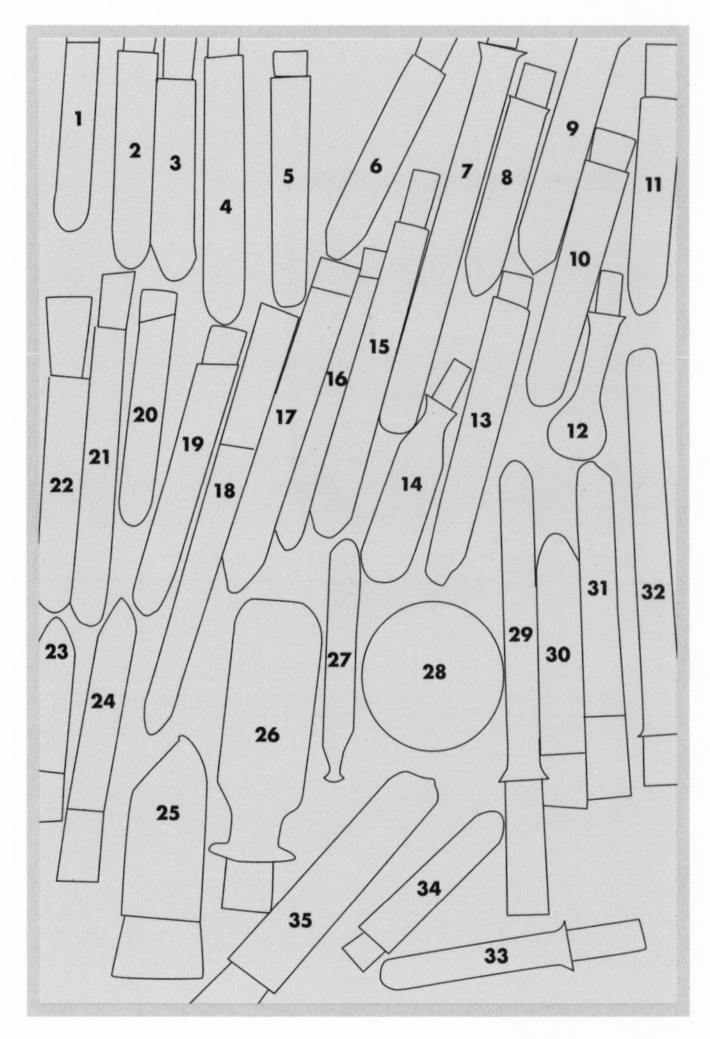

1.
2.
3.
4.
5. **Amanda Bao, age 17, enthusiast**
I've always been what you call a short person and I thought that was just something I would have to get used to. But then cold fusion came knocking, and the difference is amazing. Nobody's asked me about daytime weepings for months. It's so helpful I almost told my friends about it.
6.
7.
8.
9.
10.
11.
12.
13.
14. **Jerry McConnell, age 42, surveyor**
My friends would use cold fusion from time to time in the past but I always had my doubts. Recently cold fusion came out in a new way that squashed all those doubts. I can't believe I'm already middle-aged. That snuck up quickly.
15.
16.
17.
18.
19.
20.
21.
22.
23.
24.
25.
26. **Jorge Cachueta, age 29, surveyor**
I never thought about using cold fusion. Then again, I thought computers were a bad idea. Now I use cold fusion every day—even in the bedroom! I could try a cheaper solution, maybe even a better one. But why risk it? After all, the long-term damage is still disputed.
27.
28.
29. **Margaret Madsen, age 65, retired surveyor**
I used to make a lot excuses and pretty often they worked. But now with cold fusion I'm way more in control of myself, and others. It's a great feeling to know that others know exactly where they stand with you. I really need to get out of here soon. I hope I still know the way.
30.
31.

THE BENNY ZONE
Hi, this is Benny.
I live in the same house as Dr. Doris. Welcome to Benny's Zone! This is my zone to do whatever I want in, just like the shed back behind our house.
I've got better things to talk about than cold fuchsia. Here you have some of my best facts and findings from my life and also things I've learned.
Let's get going!... To my next page. Here it comes.
BENNY

WHICH OF THESE MEN SLEEPS IN MY HOUSE?

CRACKERS

There's always one thing in every house that runs out first. In our house it's crackers. Crackers are great and they're also as fun to eat as to say. I can eat them at all times of the day, even at 3P.M. I like the way they taste and also their flavor in my mouth.

Some people go to the park and feed pigeons crackers. It's crazy to feed pigeons crackers and it makes me upset. Pigeons don't know how good crackers taste. If you give them crackers knowing that they don't know that, then you're just wasting the food on them. Sure, they'll eat them. They'd even eat a pocket full of crackers each I bet. But what I know and the pigeons don't is that their brains are way too small to fit cracker enjoyment into. That means they just eat the crackers... without feeling a thing.

After lunch, I'm not allowed to leave the house again.

Match the bird to its favorite song.

BENNY'S GUIDE TO WHAT TO DO WHEN YOU'VE BEEN LEFT BEHIND

STEP 1

Try to find out where you've been left. If someone's around try to figure out how to ask them where you are. Use your hands. Point to your stomach repeatedly. If that doesn't work, scream out questions like "Washington?" If there's no answer you're probably not in Washington.

STEP 2

See if you brought snacks. A good place to check for snacks is your pocket. If you have a snack pouch look inside it. Does it have any snacks in it? If you find anything make sure and eat it all right away. You could be here for a while.

STEP 3

Dancing. Dancing is liked by everyone on earth so maybe someone will see you dancing and like it.

STEP 4

Take out your bouncy ball. This is a good time to get better at bouncing your ball. First throw it up in the air. Next let it bounce. Now try to catch after only six bounces. After you've mastered that try to go down from six bounces to five bounces. Five bounces is really good. Don't let any kids take your ball from you.

STEP 5

Boat? If you're on a boat, don't get off it. No matter what anyone or anything tells you, it will not be quicker getting home off the boat.

STEP 6

Make sure you're not in your house. Does it smell like your house? Is Dr. Doris staring at you angrily? Are you in your bathtub? Can Dr. Doris help you out of the bathtub? Maybe you are home.

MY LEAST FAVORITE TYPES OF WATER

OLD WATER, UGLY WATER, SEA WATER, SALT WATER, HEAVY WATER, LITTLE WATER, SCENTED WATER, BOTTLED.

MITTENS vs SOCKS

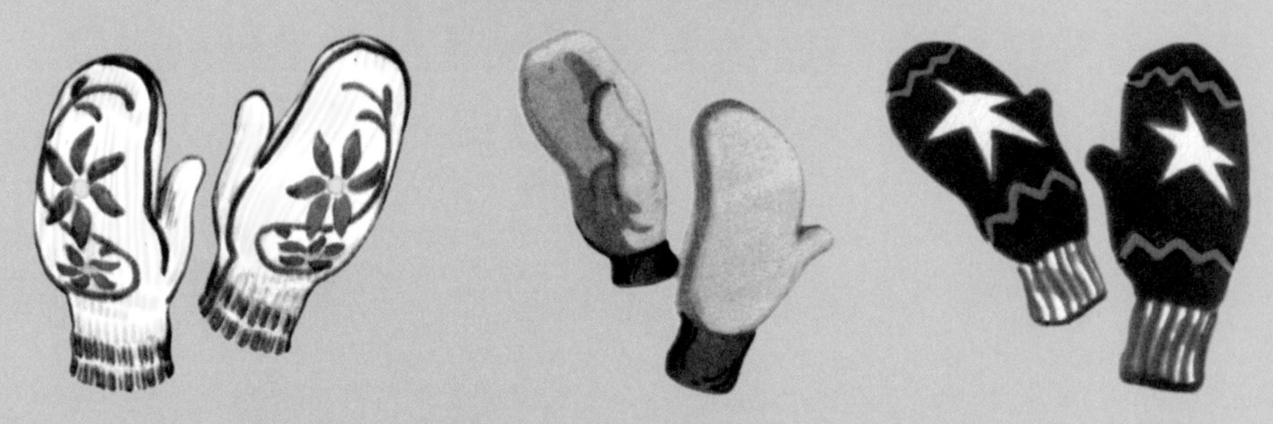

Socks and mittens are definitely the best but which is the one that's definitely the best? Good question. I like mittens. Wait. Socks. Yeah, socks.

I usually put socks on my feet so that makes them different than mittens which you can put on anywhere. But socks come out of the dryer hot and then are great for face rubbing. I tried making snowballs with socks. That's hard. Snow is for mittens. Mittens and snow angels. And I make big angels too. Dr. Doris says so.

Mittens only have one finger hole and another hole for your hand. Sometimes I put my thumb in the hand hole. But then my other fingers get confused. Usually if I wear one sock on one foot I end up putting something else on the other foot. But socks hate water more. Especially baths. But socks come in over four different colors whereas my mittens are just green.

I need to go. My legs are asleep because I left them under the heavy boxes.

RADIOACTIVITY

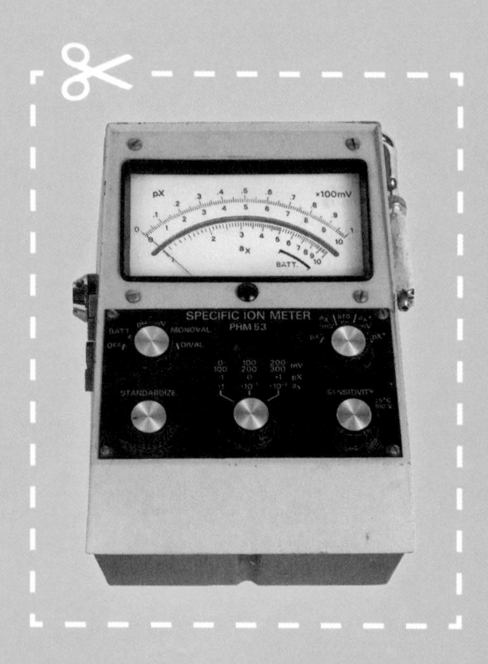

Careful! Use safety scissors and a bib.

Look here! Till you get dizzy.

HIDDEN ANIMALS

Q: How many animals can you find in the trees?
A: I think they are dead.

TONGUE TWISTERS

Potted Plant

Bagel Sandwich

Pants

Apple Carrots Apple

Europe

Blueberry Muffin Pancake Bread Sauce

Bank Statement

WHY POTATO CHIPS ARE DELICIOUS

REASON 1. Their taste. When you eat them they taste good and when they are not in your mouth you try to put them in there so you can taste them.

REASON 2. Their shape. They are shaped kind of round, which is good, but they also have jagged edges and that's why dogs don't eat them unless they're on the ground.

REASON 3. They are not squash. If they were squash then they would be probably living in the garbage. But because they are chips they're eaten and form a smile in your stomach.

REASON 4. They have to be eaten in trees or under the house. This makes it fun cause then they're a secret food which are the best kind, and also if you eat them there Dr. Doris keeps giving you an allowance.

REASON 5. Chips is the third best word. After the words "pool" and "slide," chips is the best word to say and to talk about and it also serves as a great source of flat potatoes.

PAINT BY ATOMIC NUMBER

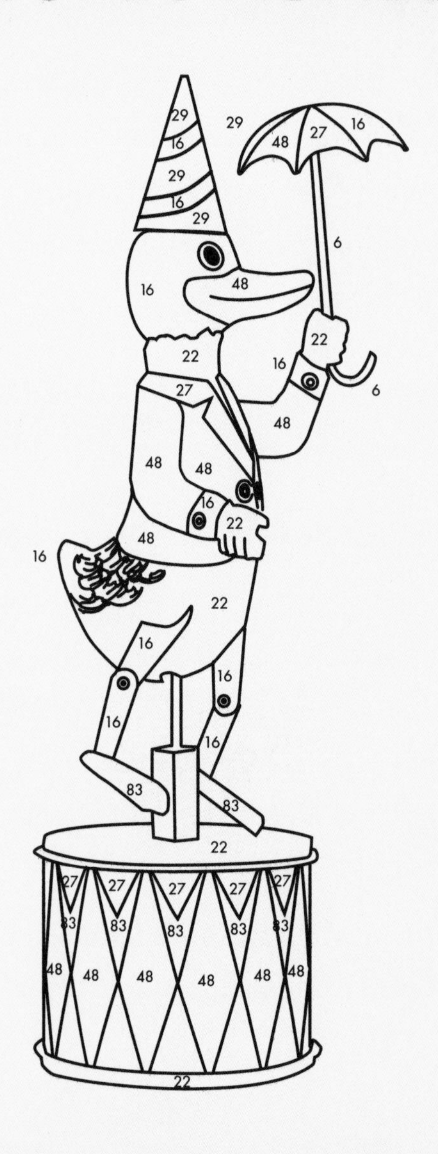

NOTES FOR THE MIDDLE OF THIS BOOK

First of all. Stop underlining my book. Not only is it a disgusting habit but you are also implying that certain of my sentences are more important than others. Where do you think you are? This isn't McGraw-Hill. There are no filler sentences or patter within these foiled covers. Every line is as finely crafted as you will find in American literature, never mind a scientific annual. Save your highlighter for lesser books.

And please stop flipping around the pages. They're ordered precisely and deliberately to maximize your chances of actually retaining something.

No more headphones. Turn off the radio in the next room, too. The information here is too powerful to be diluted with nonsense music like NWA or Chopin.

Stop buying the Professor Rumple-Feathers Haggis-On-Whey Companion Series books. He is no companion of mine or this series. He's just some halfwit with a funny name.

No snacking. Don't smudge my book.

Don't read Benny's bits. That only encourages him.

Scan the notes you've been taking and let me see them. Don't be embarrassed. I already know you're doing something wrong but I just want to know what. Send them to the address at the bottom of this page.

PERSONS & THINGS WELL-KNOWN FOR BEING MIDDLING

Alice Nelson
Ham
Leviticus
Lisa "Left Eye" Lopes
Crackle
Lebanon, Kansas
Beverly Hills Cop II
Andorra
Agnostics

Quit reading outside. Would you bring a World Almanac to the park with you? Stay inside where you won't harm anyone.

IF YOU ENJOY H-O-W...

You will likely take pleasure in Prof-R-F's

"JOY OF TOMES"

series of books with similar titles

Welcome, a thousand times welcome!

Yes, young faithful. You have found me and thus I you. Here in me and my series you have found a partner in your voracious appetite for all things Haggis-On-Whey-like. If you like their books, you will love mine. Or perhaps simply *like* mine. It matters not to me just so long as you are happy.

My name is Professor Gary Rumple-Feathers and my books are filled with graphs and analysis, stories, extrapolations, reimaginations, pontifications and the same puzzling enigmas that you've come to expect from the Haggis-On-Wheys. My books are all titled similarly—usually beginning with the words "Under a"—and all are laced with subliminal Christian or Pagan themes. Enjoy!

For the voracious book-devouring child who puts down the encyclopedia or Bible or wiccan manual and cries, "More, more!" I have provided additional refuge within my pages. And now I graciously invite you to join me aboard my leather-bound ship into the seas of worldly whimsical wonder.

Signed,

Professor Rumple-Feathers

Prof. Rumple-Feathers

THESE BOOKS ARE NOT READ OR ENDORSED BY DR. DORIS HAGGIS-ON-WHEY. DR. RUMPLE-FEATHERS IS THE NEPHEW OF OUR COPY EDITOR.

THE STORY OF BUNSEN AND HIS BURNER

Brad.

As you would assume, Robert Bunsen, creator of the Bunsen burner, was a small-time exotic-animal smuggler. But like many in the field, he failed to provide an adequate living for himself. So he turned to smelting. And Bunsen was a happy smelter. Happy and proud of his great proficiency at melting and fusing ores in order to separate their metallic constituents. But even so, he still felt something lacking—a lack that went beyond his shortage of any adult-sized organs. That is because Bunsen had made no mark on the world. Sure, he had dabbled in poetry and had made many false paleontology claims, but he just couldn't seem to find something that would stick. So he started spending all free time in the local laboratory, determined to make his mark in science, his one true love. For three years he spent every spare hour obsessed with making a discovery. But after three years, he had discovered nothing. All appeared lost. But there is always hope around the darkness near the dawn of tomorrow's darkest hope of tomorrow's night. And so one day, all his tireless work and sacrifice finally paid off. On this day—wait, it was actually night, nearly midnight—Bunsen was at the lab along with another local scientist, whose name was, I think, Brad. At the stroke of twelve, Brad left the lab pleased beyond words after having finally finishing the prototype for a brilliant new invention that could produce a hot, sootless, non-luminous flame with a simple, economical handheld metallic device. Bunsen wasted no time at all. He grabbed Brad's brilliant new burner and set off into the night. Ten months later, after a whirlwind tour through America and Mississippi, Bunsen had made his mark. He was celebrated everywhere for making countless worthwhile and totally worthless scientific experiments possible in laboratories and garages all over the world. For generations to come, scientists, bad scientists, and reluctant high school chemistry students would know his name, and in some cases build small shrines to him in their foyers. And what of Brad? Brad was okay. Brad was very good-looking, so he was okay.

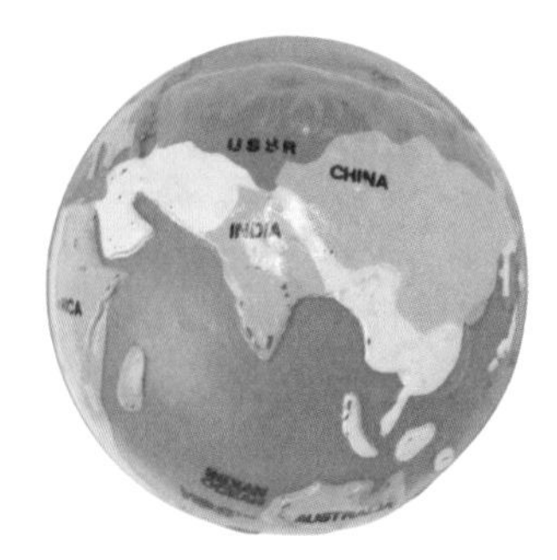

Before Bunsen made his mark

After Bunsen made his mark

After the giant red-shirt people appeared

MAKE YOUR OWN TEMPORARY TATTOOS

Everybody loves a great temporary tattoo. They feel great and look appealing and they're a great way to give people insight into your favorite objects or mythical creatures without having to actually communicate with them. Now follow these steps.

INSTRUCTIONS

1. First, pick one of the following seven drawings pictured at right you like the best to be your new tattoo (temporary).
2. Pick the destination on your body for your new piece of body artwork.
3. Using a scanner, a rear projection device, and a 3D printer, transfer your drawing to a sterile piece of transparency.
4. Send this piece of transparency to one of the following temporary tattoo transfer labs in Estonia along with a self-addressed envelope:

   Bilov and Bezzaborkin's Fake Body Skin Inc.
   Peterburi tee 81
   Tallinn, Estonia EE–11415

   Tätoveering Triumfeerima
   Weizenbergi 37
   Valge 1 Tallinn, Estonia EE–10127

   Jõgela Küla Eesti Ühispank Infamous
   Karu St. 16
   Tallinn, Estonia EE–10120
5. Set your timer for two weeks.
6. Thoroughly wash the intended body part where your tattoo will reside. Keep this area sterile. Maybe even wash other areas on your body to be sure.
7. Open your mailbox and retrieve your new temporary tattoo. Cut it out and peel off the plastic covering.
8. Put the tattoo on your body. Light candles and call friends.
9. Crank up the Bruce Hornsby.
10. Crank it even louder. That man can play the ivories.

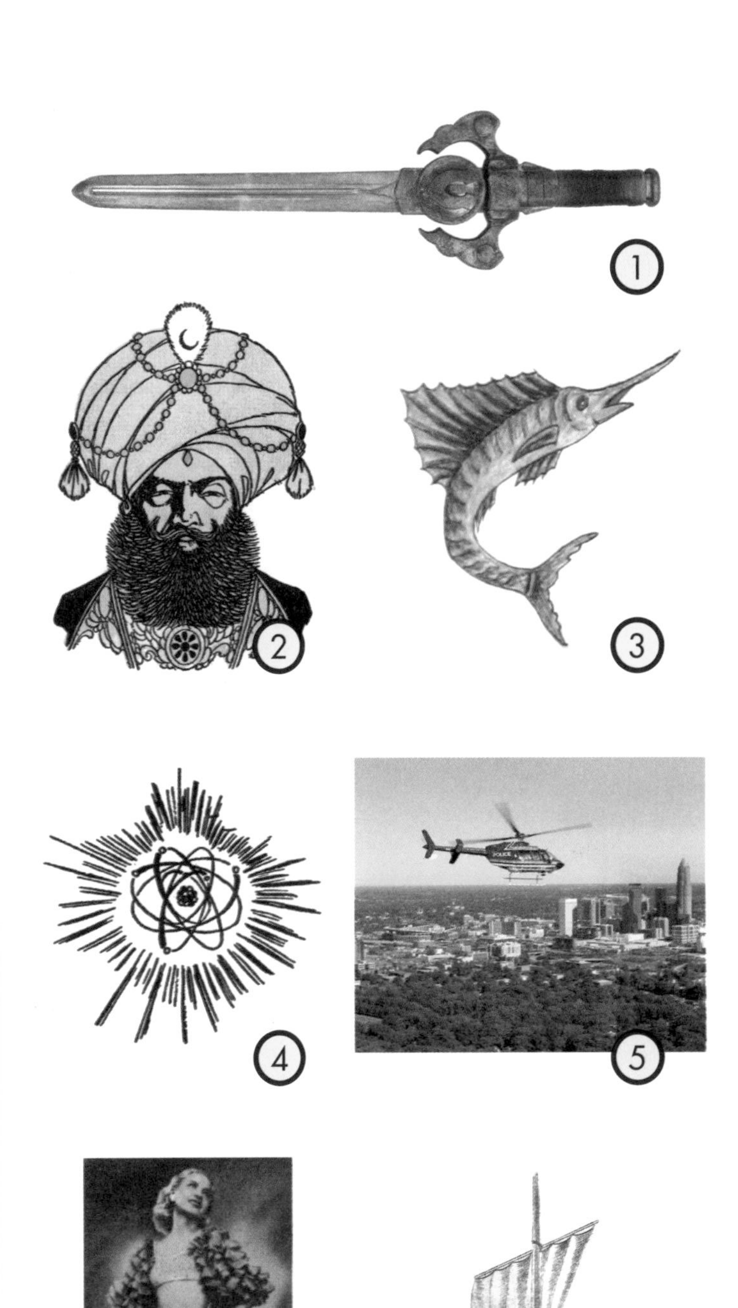

COLD FUSION CREATION ATTEMPTS THROUGHOUT HISTORY

ATTEMPT #4

Reginald Blackenwhiter

Delaware: Friday, July 2nd, 1974

In the summer of 1974, Reginald Blackenwhiter was at home waiting for his suit to be starched. He had previously been in the dry cleaner's waiting for this suit to be starched but the store had lacked chairs. So he went home, and soon Reginald was at home lying on his wooden floor watching the fan above spin because its quick revolutions felt like a more satisfactory indicator of the progression of time than a clock.

When the fan lost its hold on his attention, Reginald shifted gears and picked up two deuterium isotopes (as he did on occasion) and began playing with them, with no particular plan in mind. But as often happens on Sundays, things quickly escalated.

By the end of the afternoon, Reginald was knee-deep in sub-nuclear byproduct and had spent four hours attempting to overcome the Coulomb barrier. He looked up at his clock. It read 4:48. The dry cleaner was closing in twelve minutes! Reginald quickly brushed the quartz off his pants and ran out the door, greatly anticipating those well-starched pants. He ran into the store, snatched the pants from the clerk's hands, and paid his $14.96. He was so excited to wear these pants that he changed into them then and there—in the dry cleaner's.

Reginald's namesake, Reginald Musgrove.

Of course, had Reginald stayed another six minutes in his home and completed his tinkering he would have become the first man to ever reproduce the theoretical effects of cold fusion in room temperature. But then again, it's hard to overemphasize just how great those freshly pleated slacks looked on him. They looked amazing and felt even better.

ATTEMPT #23

MARK JEFFERTON

ALABAMA: SUNDAY, THE 12TH OF AUGUST, 1954

Mark had been left alone in his house for the weekend while his parents set out to become the first couple in Alabamian history to sell pre-manicured fingernails wholesale. Mark was in the hallway, quietly sucking down yogurt as he was apt to do when bored or nervous, when out of the corner of his eye he saw a 250ml cylinder sticking out of the closet of his parents' room.

He knew he was not supposed to mettle in his parents' closet affairs but something about the roundness of the glass compelled him. He had to hold it in his hands. He rushed into the room and seized the perfect cylinder. It was cool to the touch and light in his palm. This was truly a marvelous beaker. Mark snuck away, back to his room with the beaker in hand.

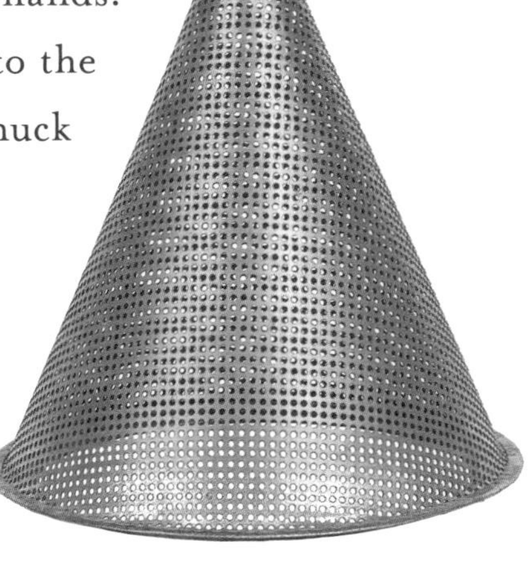

A COLANDER.

He laid the beaker on his nightstand and as he did so Mark thought he heard a muffled voice talk to him. He thought better of it, got out his magnifying glass, and put it to the wonderfully crafted vessel. But the voice came again and now he was sure of it: the beaker was talking to him.

"Mix me," said the beaker. "Mix me!" it demanded.

Well, Mark was an impressionable young man, so before you could say Berkelium, he had run off to gather the isotopes from his mother's secret locked drawer and the colanders from his father's tool box. He even found grandpa's old spectrophotometer. He laid them all before him.

The beaker barked at Mark, pressuring him into more and more scientific advances until he was mere seconds away from forcing two light nuclei to fuse and expel a massive doze of energy upon the room. But just then, Mark's neighbor Mrs. Muglia walked in behind him and caught Mark red-handed. "Mark!" she cried. And with that, the jig was up. *Continued...*

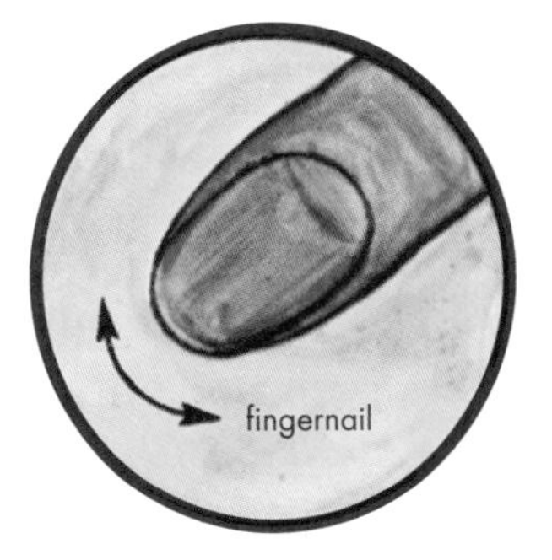

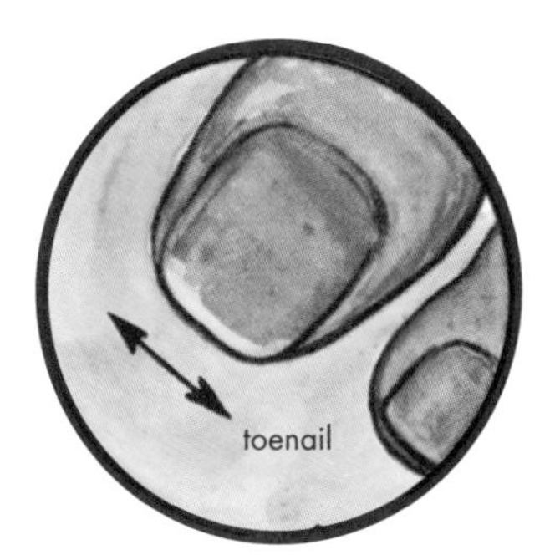

FOUR PICTURES OF FINGERS IN CIRCLES.

ATTEMPT #12

President Taft

Cincinnati: 1868

William Howard Taft was not yet president. In fact he was a small, fat-cheeked boy currently pinned down under the legs of Vance Potsdam. Because it was Friday there would be a second round of welts delivered to William's thighs and arms, enough to last the weekend. This had been carefully explained upon their first meeting, when a foresight-lacking William Howard put his poorly defined face right in the way of Vance's conversation about the newly invented fabric rayon.

The Taft family summer home.

After twenty minutes, Vance determined that a weekend's worth of bruises had been sufficiently administered and thus left to go play a round of Mumblety-peg with his friends. Meanwhile, the thoroughly bruised future twenty-seventh president got up, wiped off his Fauntleroy trousers, and ran home.

Molasses was a recent invention and one of which Billy Taft was a big backer. Once home, he wasted no time in climbing up to the high snack-laden cupboard and grabbing the molasses jar hidden behind the many bags of flour. Without looking, he set the jar on the table and prepared to eat. But his greedy, fat hand encountered a problem.

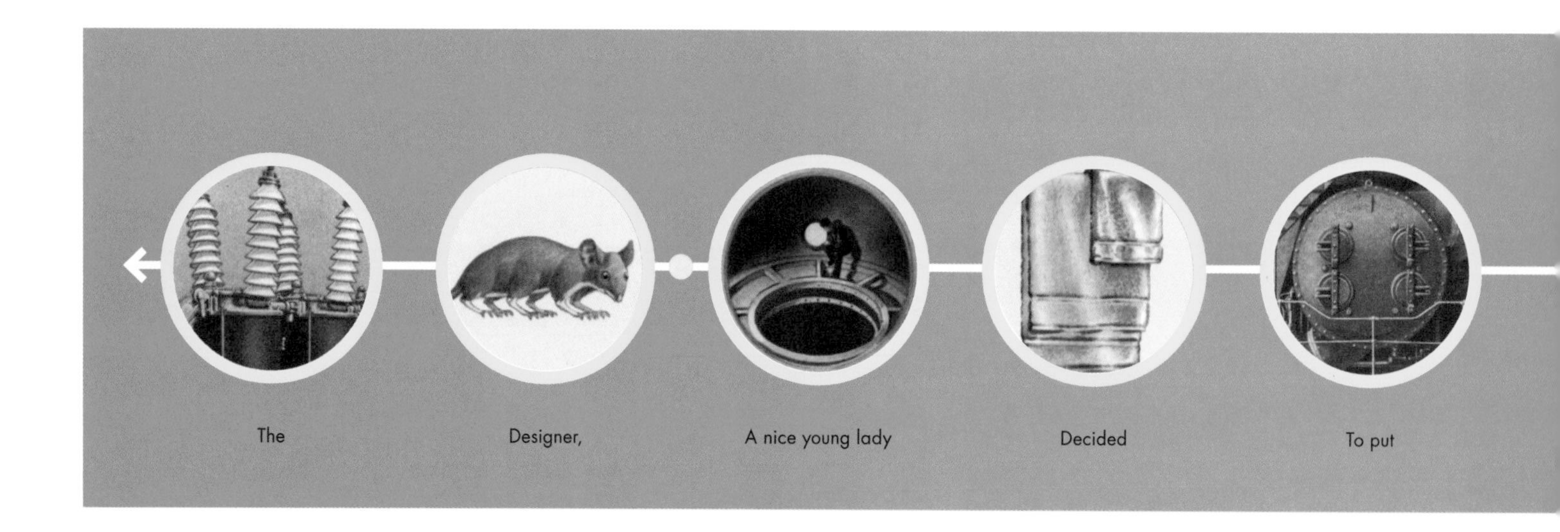

THE TAFT FAMILY DAGUERREOTYPE.

The jar was bone dry. Some other member of the Taft clan—from his portly sister to his well-insulated father—must have cleaned it out. Covered in welts and devoid of sugary comfort, Tubby Taft leapt into a whirlwind of rage. He kicked over his can of jackstraws, laid waste to his Bilbo Catcher, and put a swift end to the family daguerreotype. Along the way Taft dismantled his mother's hand-painted loom and rendered his sister's zoetrope only a memory. During the melee, the commotion sent charged copper shards (from a freshly destroyed motor sewing machine) flying into the molasses-vapor-heavy jar. Had the tender drumstick of a boy simply allowed the charged electrodes a few brief more moments to fester and reach fruition, then that Friday might have been the first day in cold fusion history.

But as his campaign slogan of 1910 "If you think it, do it" later revealed, William H. Taft was not a champion of patience. It was this same insolent nature that led the Spanish king to call him "El Tirón." And so Taft, with his doughy arms as high as gravity would allow, his hands firmly clutching the family iron cauldron, let metal meet glass with a scientific progress-halting thrust, smashing the large jar into uncountable pieces. With nothing in the house left to break, the still-peeved William Howard Taft set off to the woods with a plan to beat a set of trees mercilessly with their own fallen branches. As he did so he smiled, knowing he would someday become one of the country's worst presidents by far.

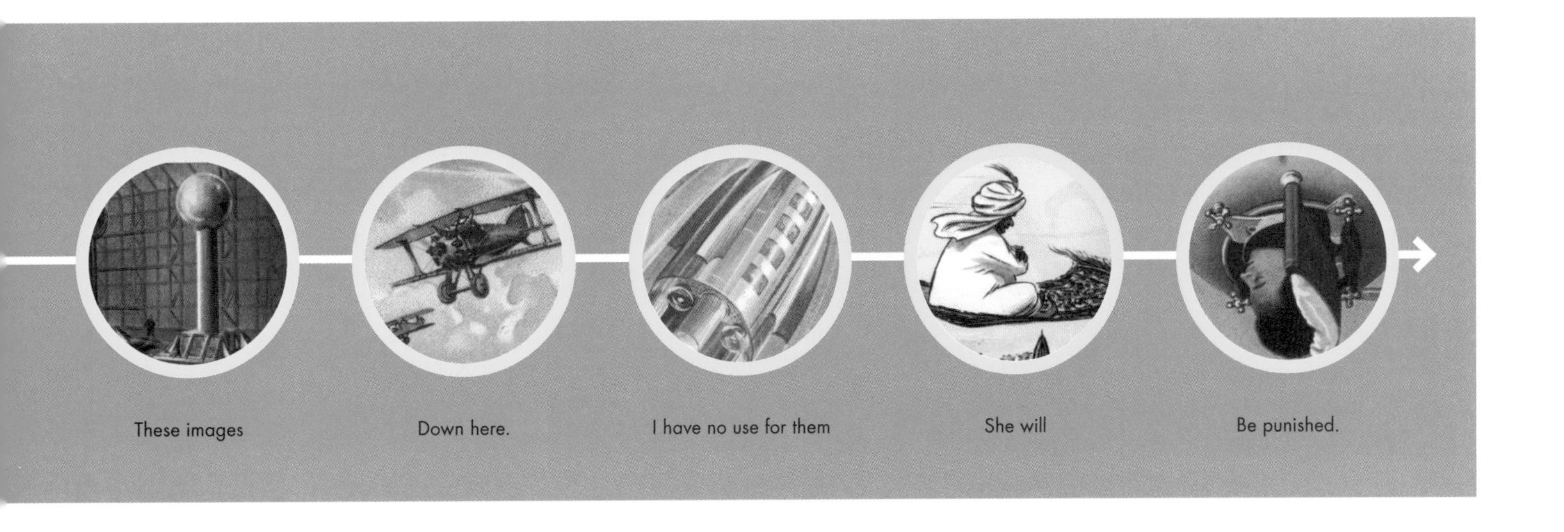

HOW DID THE NOBLE GASES BECOME NOBLE?

There was a time, following some of the bad-boy antics of oxygen and hydrogen, when we at the World Science Council, which I invented, started to wonder if any gases could still earn our trust. They had become devious and scummy like a snake with an abacus.

But, it was at that moment—through self-sacrifice and acts of courage and valor—that six gases rose above their peers to once again gain the trust of human scientists. These became known as the noble gases. These are their stories...

HELIUM
Bravely volunteered to become the shielding gas for use in most arc-welding processes.

ARGON
Heroically became the dominant gas for use in extinguishing fires involving server equipment.

NEON
Daringly found its place as the commercially used cryogenic refrigerant in applications not requiring the lower temperature range attainable with more expensive liquid helium refrigeration.

RADON
Gallantly stopped killing people by removing itself from their building materials and indoor facilities.

KRYPTON
Selflessly lends its powers to flashes for high-speed photography.

XENON
Courageously switched from use as a dimmer in excimer lasers to use as the primary fuel component for ion propulsion of spacecraft.

THE TOP ELEMENTS OF 2009

Element	Overall rank	Cumulative total	Poise	Charm	Grooming	Evening gown	Community service	Academics	Student body	Campus life	Half-life	Slugging pct	On-base avg	WHIP	ERA	Khakis	Viscosity	Madeline Khan	Wrath of Khan	Balsa
ERBIUM	**1**	**303**	5	8	5	10	80	3	47	9	3	8	8	61	23	8	10	4	4	7
MOLYBDENUM	**2**	**178.3**	10	10	10	9	10	10	10	10	10	10	9.3	10	10	10	10	10	10	10
SEABORGIUM	**3**	**153**	10	8	10	3	9	4	8	10	10	10	9	9	9	9	9	6	10	10
PHOSPHOROUS	**4**	**146**	8	10	8	8	8	8	8	8	8	8	8	8	8	8	8	8	8	8
BROMINE	**5**	**142**	10	9	9	2	10	9	0	9	4	8	9	6	10	10	10	10	13	4
HYDROGEN	**6**	**135**	4	7	6	4	9	7	4	6	3	0	16	17	4	N/A	6	33	9	0
BERKELIUM	**7**	**130**	8	8	10	9	4	7	7	7	7	8	4	5	3	9	9	8	10	7
TECHNETIUM	**8**	**121.5**	3.3	6	9	6	8	7	5	5	20	4	9	8	3	7	2	9.2	6	4
RUBIDIUM	**9**	**120.5**	7	9	4	5	9	5.5	10	9	10	10	10	7	4	4	7	6	2	2
BORON	**10**	**119.5**	1	2	6	7	4	4	7	10	10	10	9	10	5.5	9	5	4	9	7
TITANIUM	**11**	**117**	3	9	8	7	7	22	6	4	5	7	5	3	4	8	6	5	1	7
OSMIUM	**12**	**110**	7	7	7	7	7	6	5	7	4	7	6	9	7	8	5	6	3	2
XENON	**13**	**90**	5	5	5	5	5	5	5	5	5	5	5	5	5	5	5	5	5	5
ALUMINUM	**14**	**89**	9	4	-2	9	4	7	4	10	3	7	5	6	4	2	10	3	0	4
TUNGSTEN	**15**	**88**	4	3	3	10	9	5	2	1	7	4	4	9	6	4	6	4	1	6
MANGANESE	**16**	**85**	8	1	9	4	7	3	1	7	5	8	N/A	4	6	6	2	3	6	5
HAFNIUM	**17**	**84.3**	7	8	9	0	7	4	8	0.3	7	5	2	2	7	3	8	2	4	1
ZINC	**18**	**81**	1	2	6	7	2	4	9	8	9	2	N/A	3	5	4	5	6	3	5
SODIUM	**19**	**80**	5	8	1	9	3	7	4	2	6	2	8	5	2	1	4	8	2	3
PRASEODYMIUM	**20**	**76**	3	5	0	4	2	4	5	1	3	9	N/A	3	8	1	8	3	8	9
PLATINUM	**21**	**74**	2	9	4	8	8	0	4	6	8	2	N/A	1	4	9	3	0	4	2
EINSTEINIUM	**22**	**73**	3	1	4	5	5	3	5	4	6	3	N/A	6	5	8	2	1	3	9
ARSENIC	**23**	**70**	7	9	2	1	3	9	2	4	5	5	2	5	2	5	3	5	0	1
GERMANIUM	**24**	**70**	4	4	2	1	5	3	4	6	5	1	N/A	4	8	3	4	4	9	3
COBALT	**25**	**69.3**	1	8	3	0.3	2	3	1	6	2	9	8	9	5	6	1	2	3	0
IRIDIUM	**26**	**67**	8	5	0	2	5	6	5	0	5	4	N/A	0	2	7	4	7	3	4
LUTETIUM	**27**	**66**	7	0	7	1	2	1	7	3	5	4	N/A	7	3	6	1	4	7	1
HELIUM	**28**	**66**	8	4	3	6	2	1	7	5	3	3	N/A	0	4	1	8	2	6	3
IODINE	**29**	**65**	4	2	9	5	0	9	0	5	3	2	N/A	3	7	1	5	3	4	3
IRIDIUM	**30**	**60**	4	5	0	4	4.3	8	7	3	1	3	N/A	6	3	1	6	1	3	1
KRYPTON	**31**	**58**	5	2	2	6	4	8	3	3	5	4	N/A	1	3	1	2	4	2	3
TANTALUM	**32**	**46**	2	0	1	7	2	0	4	6	4	3	N/A	2	1	1	6	3	1	3
THULIUM	**33**	**44**	1	3	NA	9	5	2	0	2	5	1	N/A	2	0	0	3	2	7	2
PALLADIUM	**34**	**42.12**	1	1	2	3	5	8	0.13	0.21	0.34	0.55	N/A	0.89	8	5	3	2	1	1

NOBLE GAS **TRANSITION METAL** **METALLOID** **HALOGEN** **OTHER**

An early fruit fusion experiment, which laid the theoretical groundwork for Dr. Pachovonich's breakthrough. Note that the number of small white lines emanating from the two fruits is *exactly the same*.

THE RUSSIANS AND THE INVENTION OF FRUIT FUSION

In the 1960s, as the Dietary War raged on, thousands of miles away the Russians found themselves grossly outmatched in the fruit category. While the variety of resources was obviously direly low compared to their American opponents', the Russian culinary theorists had the edge in creativity, as evidenced by the Fedoseev/Andravanich Cobb Salad variation first introduced in '57. While the combination of two to three fruits had been more or less mastered at this point, the concept of a true fruit medley comprising seven to eight different fruits still seemed the theory of dreams and fantasy. That all changed on the night of October 23, 1968. While working late one night trying to come up with a way to blend melons without symptomatic flavor assimilation, a Dr. Pachovonich discovered a principle that would shatter all previous fruit immersion theories that had existed in the annals of food thought: fruits are succulent and delicious. Thus the good doctor began combining fruits at a breakneck pace and seemingly at random. His colleagues grew apart from him, fearing his blending to be reckless and possibly dangerous to their own careers. Meanwhile *Easy Rider* was really making a stir in Hollywood. Peter Fonda was truly a sight to behold. People were flocking to see the film and were enjoying it.

SUBJECTS FOR DOCTOR

	DEMOGORALS	REINSTITUTES	TEMPORALS	SYMBIOS
GERIATRIC				
HYPERBOLIC				
MESOPOTAMIC				
HYPEREXTENSIVE				
VOCATIONAL				

DORIS'S FUTURE INQUIRY

JUDICIANARIES LATENTS JUBILANTRIES FRACASES

WHY BIRDS ARE BAD AT BUILDING SUPERCONDUCTORS

When the general public thinks of palladium and heavy water they often think of birds. "Why not have birds do the lab work?" they ask. "After all, they can fly, and sometimes look pretty." So they hand Dewar flasks and calorimeters to any winged creature and think the rest will take care of itself. But we must remember the limitations of birds, and their inherent flaws. Birds, for example, do not use commodes. They just drop their feces on the ground while flying; they don't even wipe. So that's one thing.

Site of attempt by some birds to build a superconductor in Harrisburg, PA. They failed.

Site of attempt by some other birds to build a superconductor on the "Big Island" of Hawai'i. It really was kind of laughable.

Site of attempt by some birds to do some kind of nuclear activity in Deerfield, IL. They didn't even have a Dewar flask.

Site of attempt by some birds — a few of the same ones from Harrisburg — to build a superconductor outside East Gardiner, OR. This one was actually pretty good.

BENNY'S BOOK SUMMARY

I just looked through this entire book. It doesn't seem to be about faces or holes at all. Instead there's just a lot of words, science tubes, and things not covered in fur.

This makes me feel like I feel when I build a stick castle too close to the river just before it rains. Doris has changed some of the locks inside our house again.

Cats can jump really high but most of them don't. Dogs jump a lot more but not as high. I never jump but probably when I do it's going to be high. I'm saving my jumping like it's a red Skittle.

This book has between 256 and 266 pictures in it. I've touched all of them and the ugly ones feel the same as the pretty ones.

Tomorrow we're going for a drive and that's good because I know where I'm going to sit. Even when I close my eyes I can tell where foods are being eaten nearby, just by sniffing.

It's really hard to make towels look like the ocean. And even if you do, you can never make them as wet.

I need to go. I'm tired from either hopping too much or not enough.

Enjoy the rest of your time in this book.

BENNY

Benny

COMPLAINT CARD

Your name ______

Address ______

Home phone number ______

Employer and supervisor's name ______

Work address and work phone number ______

Social Security number ______

Names of loved ones ______

Addresses of loved ones ______

Work Addresses of loved ones ______

Phone numbers of loved ones ______

Businesses you frequent ______

Make and model of car; license plate number ______

Travel route to and from work ______

Weekend routine ______

Holiday travel routine ______

Names of friends with criminal records ______

Their addresses and phone numbers ______

Their sites of incarceration ______

Their parole conditions and the results of their hearings ______

Their measurements and allergies ______

Complaint ______

UPCOMING TITLES FROM THE H-O-W SERIES*

Pews: History and Lore

Prince Charles, King of Kings

Managing Water Distillation Plants

Reinforced Concrete Mixing

Church Bell Testing and Scoring

Symptoms of Nitric Acid Inhalation

Quarry Construction:
Profiting Under the Watson Clause of the Title 27

Federal Mandates and Avoidance

Making Faces at the Elderly

Horse Stable Architecture

Plankton

Plankton II: The Green Army

Lime Mortar Installation and Maintenance

Different Kinds of Fences

Zinc Extraction

Long Distances

What the World is Jonesing for Now

Placards and Placation

Extreme Action Sport Athletes
and the Accountants Who Love Them

Felt Hats

Sturdy Felt Hats

The Heideckers of Wisconsin: Truth and Myth

** subject to change and become better*

ABOUT THE AUTHORS

Dr. Doris Haggis-On-Whey has thirty-two degrees from thirty-one universities and colleges, some of them in England. She has circumnavigated the globe at least four times, and has traveled widely. She has or will eventually author 176 books about subjects ranging from the Earth, its animals, giraffes, and melanoma. She has been awarded awards, including the Prize Award, which is given biannually to the scientist who has won the most awards in the previous eighteen months. She lives in Crumpets-Under-Kilt, Scotland, with no pets and Benny. Benny is smelling the headlights on the truck down the road.

ABOUT THE DESIGNER

Eliana Stein is honored to have followed in the footsteps of Mark Wasserman and Irene Ng. She would like to reassure the reader that Dr. Doris demonstrated restraint and compassion in meting out all punishments received for work on page 47.

Research assistance has been provided by the de la Manzana brothers, and by Toph and Dave Eggers.